Volney Rattan

Excercises in Botany for the Pacific States

Volney Rattan

Excercises in Botany for the Pacific States

ISBN/EAN: 9783337106188

Printed in Europe, USA, Canada, Australia, Japan

Cover: Foto ©berggeist007 / pixelio.de

More available books at **www.hansebooks.com**

EXERCISES IN BOTANY

- FOR THE -

PACIFIC STATES

- BY -

VOLNEY RATTAN

TEACHER OF BOTANY

IN THE

CALIFORNIA STATE NORMAL SCHOOL

SAN JOSE

THE WHITAKER & RAY CO.

SAN FRANCISCO

- 1897 -

PREFACE.

This book is specially designed to supply the needs of pupils who must work with simple appliances at ordinary school desks, and under the supervision of a teacher who can devote but little time to the subject. But it is believed that it will be equally useful to those who have the advantages of a well-equipped laboratory and the aid of a special teacher.

An attempt has been made to lay out the work from day to day, so that the teacher, burdened with other duties, need have little to do in the way of preparing outlines of the daily work. It is expected that the pupils, not the teacher, will provide all the material used. Each exercise directs work easy enough to be profitable to the weakest pupil in the higher grammar grades, and at the same time suggests problems which will try the power of the brightest pupil in the last year of the high school. The object of the exercises is chiefly to teach how to study plants, not to give information about them. A few facts are given for the purpose of encouraging pupils to look for more of the same kind, and, at the same time, furnishing material for the important work of verifying the discoveries of others, by repeating the observations or experiments which revealed them. President Jordan says, "To verify the fact gives training; to discover it gives inspiration. Training and inspiration, not the facts themselves, are the justification of science-teaching." The ability to get facts and to use them, not the number of facts acquired, is the true measure of education. The value of observed facts to the student, like the use of minnows to the fisherman, is that, if skillfully handled, they enable him to get larger facts. It is this exercise of handling facts that strengthens the mind, that truly educates. As the activity of muscles and mind is worth more to the hunter than the game he brings down, so the exercise of observing facts and deducing truths from them is more valuable to the student than the knowledge gained, or, as President Jordan puts it, "To seek knowledge is better than to have knowledge." The wise teacher does not, in giving an educational test, direct the student to write out what he has read; but says to him, "Go to-day into the library, the laboratory, or the fields and find facts; to-morrow tell me what they mean."

The questions in this book are designed to draw attention to facts that might otherwise be overlooked; to stimulate thought and to lead in the direction of truth. They are not test questions. It is not expected that pupils can at once answer many of them, nor is it supposed that any pupil can, even at the close of the course in botany, answer perfectly all of them. The teacher should never answer any of the questions. They are for the use of the pupil. He should, however, help the class to weigh the evidence presented by pupils in defense of their answers. As a presiding judge he should point out errors in statements of facts and weak points in arguments. The questions must be decided, if at all, by evidence, not by authority. It must be remembered that queries have the most educational value to one who reasons out the answers from facts which he himself has discovered. Once more quoting the highest authority on science-teaching: "The purpose of science-teaching as a part of general education is this—to train the judgment through its exercise on first-hand knowledge."

Since in most schools botany is taken up soon after the winter holidays, the first exercises are devoted to seeds and their germination. If the work is begun in August or September, flowers, fruits, and leaves should be studied first, then seeds and winter buds. In the study of leaves, flowers, winterbuds, and entire plants the number of exercises can be extended indefinitely. Indeed, new exercises may be intercalated, or those in the book omitted, at the discretion of the teacher. Good material should be used when it can be easiest obtained. Two days may sometimes be devoted to one exercise. Pupils who complete the exercises on leaves, storage stems, flowers, and inflorescence will be able to use the "Key to West Coast Botany," or "Greene's Botany of the Bay Region." But the latter book will require frequent reference to the dictionary.

Botanists will note that several interesting facts are here published for the first time.

VOLNEY RATTAN.

SAN JOSE, August, 1897.

ELEMENTARY WORK IN BOTANY.

INTRODUCTION.

THE BEGINNINGS OF PLANT LIFE.

If the first rain of the wet season is followed by warm, sunny weather, specks of green will soon appear among the dry stems of last year's weeds; and in fence corners or other eddy nooks where summer winds have drifted seeds and covered them with dust, you may find perfect mats of baby plants. With a shovel skim off a few square inches of this plant bearing soil, and carefully examine it. Except a few green needles, which you recognize as spears of grass, most of these little plants consist of white stems, each of which bears at the top a pair of green leaves. Looking sharply you may see a tiny bud between the leaves; or, in older plants, this may have in its growth developed other leaves which curiously enough are not like the first two. Searching through the shovelful of earth you may find plants in all stages of growth,

Fig. 1. *a.* Bur clover seed sprouting *b.* Same three days later. the halves of the beanlike seed having become a pair of leaves. *c.* Mustard. *e.* Mallows.

Fig. 2. Eschscholtzia seedling, showing a pair of bifid leaves (cotyledons).

from swollen and sprouting seeds to stems, which are just pushing their bowed leaf-heads into the sunlight. Here, then, is material from which you may learn how plants grow; a lesson, remember, which no text-book or schoolmaster can teach you. It will be easier, however, since most of these early wild plants come from very small seeds, to take your first lessons from plants which have made a larger growth while connected with the mother plant.* In other words, it will be better to study large seeds and their first growth (*germination*) before you work upon seedling weeds. If you begin to take lessons of plants in the latter part of the rainy season, look for sprouting seeds of fruit trees in orchards or back yards where the pits have been scattered. You must have wondered how tender sprouts can get out of such hard shells as apricot or cherry stones. If you know where there are buckeye trees, try to find the

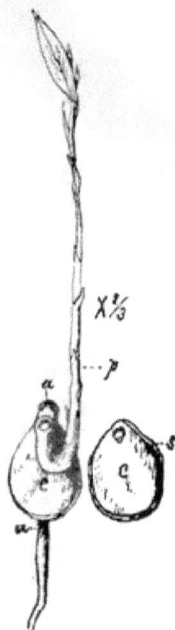

Fig. 3. Black walnut germinating, showing flat inner surface of one cotyledon. About half natural diameter

Fig. 4. California laurel. Seedling planted wrong end up, making it necessary for both sprouts to turn around. *a a.* The first sprout forming the root. *p.* The second sprout beginning to form the tree trunk. *c c.* Cotyledons. *s.* The shell, which does not split smoothly as does the walnut.

---

*The real beginning of a plant's life history is not the sprouting seed. The first exercise tells of its growth in the seed coat while fed by the mother plant. Indeed, an introduction is needed in which we are told how the mother plant prepares for the growth of seeds. [See preface.]

buckeyes—the largest seeds of any plant in our country. A baby buckeye-tree backing out of its leathery coat is an interesting object. Morever, the fact that it holds on to its coat, which is also a dinner-basket, until all the starch is eaten out is worth your attention. Acorns, walnuts, and laurel stones grow in a similar way.

Along with this out-of-door work, or preceding it by a few weeks, your most profitable work indoors will be the study of seeds, beginning with their condition before germination.* The exercises beginning on the fifth page will assist you.

*Pupils should be encouraged and urged to make as many observations as possible upon the behavior of plants under natural conditions. This work is necessary to counteract the errors which arise when we attempt to interpret experiments in which some of the conditions are abnormal. For example: seeds growing while pinned to a stick in the moist air of a closed jar do not develop just as they would in moist earth. Nor can we say without some experience what difference the abnormal conditions will make in the result.

Madia elegans.

EXERCISES

IN

STRUCTURAL AND PHYSIOLOGICAL BOTANY.

SEEDS AND THEIR GERMINATION.

Material Required.—The pupil must have at least all the seeds and implements here named:

Seeds (20 or more of each kind).—(1) Some large variety of common beans; (2) Windsor beans; (3) scarlet runner; (4) sunflower; (5) squash or pumpkin; (6) castor-bean; (7) pine-nuts; (8) buckwheat; (9) morning-glory; (10) coffee; (11) corn; (12) wheat.

Implements.—(1) A knife, which must be sharp and kept so by the use of a good whetstone;* (2) a cup or drinking glass (these may stand safely on a shelf or on window-sills between the botanical working hours); (3) a glass fruit jar holding one quart; (4) a pocket lens, if the school is not provided with dissecting microscopes; (5) a note-book.

EXERCISE 1.

Take one of the common beans. Find three marks on the edge. What do you think caused the largest mark?

---

*The teacher should have a whetstone of the kind called oilstone, on which water instead of oil should be used. Such a stone costs only ten or fifteen cents. Some important experiments require red ink, India-ink (the form known as waterproof drawing ink is best), and iodine. These should be in the teacher's charge. If the school can afford dissecting microscopes (one for each member of the class; or, better than none, half as many) they should be kept in a tight cupboard when not in use, along with needles, tweezers, dissecting knives (shoe knives answer the purpose), and other implements. All these things can be obtained of the publishers. [See Appendix.]

It is called the *hilum*. See if one of the other marks has a hole in it. That one is called the *micropyle*. Lay the bean on a piece of paper, and make close beside it a drawing natural size. Why not draw the other side? Would such a drawing show any fact not shown by your drawing? Are the sides alike as your ears, or as the buttons on a coat? This condition of sides corresponding to right and left is called bilateral symmetry. Hold your paper with the drawing on the side away from you so that a strong light shines through. Does the drawing now represent the other side of the bean? Place the

Fig. 5. Windsor bean.

bean on your paper near the drawing. Aided by an eraser, pin, or anything suitable, make the bean lie on its back with the scar up. Represent this view natural size, carefully drawing the *hilum* (scar), the *micropyle* (hole), and the *chalaza* (a double bump). Look at these marks with a lens. Take one of each of the ten kinds of seeds and place them in a row so that those most alike are together. Which have hilums but apparently no micropyles or chalazas? In which are you unable to find even a hilum? By and by when sunflowers and buckwheat are going to seed you can see for yourself that the so-called seeds of these plants are one-seeded pods. Remove the seed from one of your sunflower pods. Find the hilum. Take the seed out of a grain of buckwheat. Which is the stem end of buckwheat? Put ten of each kind of your seeds (except the wheat and corn) into a cup half full of water, and leave them there until to-morrow at the beginning of this hour. Before the next exercise hour look up in the dictionary all the words new to you.

EXERCISE 2.

Take your seeds out of the water and put them on blotting-paper or a cloth. Which have absorbed the most water? Why are the scarlet-runner beans lighter colored than they were when dry? Examine the hilum, micropyle, and chalazá of a scarlet-runner seed. Cut the skin around the edge from near the micropyle to the chalaza, and carefully slip it off. What do you find on the inside nearly opposite the micropyle? Put the kernel back in its coat. Observe the relation between the micropyle and the pocket. What fits into the pocket? The stem of the kernel is called the *radicle* or *caulicle*. Draw a side view of the kernel, and also a view with the caulicle up, showing how the two thick pieces are attached to it. Break off one of the thick pieces. Draw the piece with the caulicle flat side up. The thick pieces are *cotyledons*, and the little stem with its pair of tiny white leaves lying between them is the *plumule*. The whole kernel is called an *embryo*. Break off the remaining cotyledon and draw the caulicle with its plumule head. Remove the coats of a Windsor bean and a common bean so as to keep their pockets entire. What part of the coat is opposite the tip of the caulicle? Is it really a hole? If, when the bean is tightly covered with its coat, the caulicle should become longer it would evidently press against the bottom of the

pocket, and finally burst through, tearing either the inside
or the outside wall. Examine the pocket carefully. Which
would give way: the coat, which is the outside wall of the

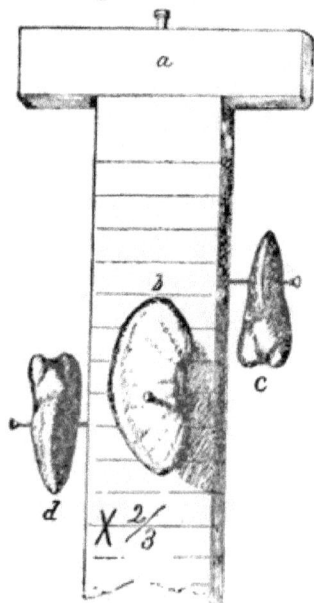

Fig. 6. Showing how beans may
be pinned to a stick for the purpose
of observing their germination in a
fruit jar. *a.* The crosspiece, which
must fit the mouth of the jar. *b.* A
Lima bean. *c* and *d.* Windsor beans,
one with the caulicle pointing downward and the other with the caulicle
pointing upward.

pocket, or the pocket wall fastened
to it? If the pocket should break
on the inside, and the caulicle
continued to grow longer, what
would be the result?

Compare the three embryos.
Examine the plumules with your
lens. Make as many drawings as
the time will permit. Put your
undissected seeds back in the cup;
pour off the water and press down
upon the seeds a wet cloth or several thicknesses of wet blotting-
paper. Prepare at home a piece of
apparatus in this way: Find or
make a smooth stick about one
inch square and half an inch
shorter than your glass jar. Cut
a piece as broad as your stick, but
not half so thick, so that it will

just fit when held horizontally in the mouth of the jar.
Fasten this with a nail to one end of the square stick, as
here shown.* It can now be made to stand up in the center

---

*The jar and stick already prepared can be furnished by the publishers.
[See Appendix.]

of the jar. Divide each side into quarter-inch spaces by
drawing pencil lines. Bring this apparatus to school
to-morrow. Look up in the dictionary out of school hours
all the new words.

EXERCISE 3.

Take the stick out of your jar. About three and one-
half inches from the lower end fasten on opposite sides by
means of pins a scarlet runner and a common bean, so that
the micropyles are at the same height and the caulicles are
pointing downward. Push the pins squarely through the
cotyledons near their apexes. On the other sides fasten two
Windsor beans; one with the caulicle pointing directly up,
the other with the caulicle pointing down. The micropyles
of these four seeds must be on a level. Place the stick in
the jar, pour in water until it touches the
lowest bean, and put on the cap air tight.
Keep the jar in a dark place or put a paper
bag over it.

Split a Windsor bean so that the caulicle
is cut in halves while the knife passes between
the cotyledons. Draw the cut side of one
half. You have represented a longitudinal,
or lengthwise, section of the seed. Could any
other section show all the parts of the embryo

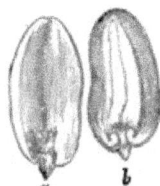

Fig. 7. Peanut
embryo. *a*. One
cotyledon at-
tached to the cau-
licle, on the upper
end of which is
the plumule. *b*
Same viewed from
the outside.

and all the important points on the coat? There could be
other longitudinal sections. Accurately described, this one
passes through the center parallel to the plane of the cotyledons. Your section must show the pocket. Put the halves

together and make a cross-section, cutting the caulicle near its tip. Cut a fresh bean the same way and draw the section. It must show the pocket, the caulicle, and the two cotyledons. Write on or beside your diagrams the names of all the parts shown, and show where the skin is thickest. The positions in the first diagram of hilum, micropyle, and chalaza should be indicated by dotted lines leading to them from their names written at one side of the diagram.

Take a sunflower seed out of its pod and remove the coat. Why is the coat so thin? Is there a pocket? Why not? Break off one cotyledon and make a drawing similar to that of the peanut on page 9.

Taste the cotyledons of the bean and the sunflower. Crush a bit of each kind on white paper. Put the unused seeds in the cup and cover as before.*

EXERCISE 4.

Which of your seeds have sprouted? What is the sprout? Where does it get out of the coat? Is the point very sharp or rounded? Draw the seed which has the longest sprout.

Study the common bean and the scarlet runner as you have the Windsor bean. Compare the plumules. How many leaves can you see on the plumule stem? How are

---

* Your seeds will behave better if now you place them on a wet cloth or paper in a pie tin or soup plate and cover with several thicknesses of wet cloth or paper; or, better still, they may be put in wet redwood sawdust or sand. The beans pinned to the stick in the jar should be those which have been soaked. It will be all the better if they have begun to sprout

the leaves folded? Do the folds of one enclose the other or
do they mutually embrace? Take a sheet of paper, fold it
crosswise and then lengthwise into four thicknesses, cut or
tear the side opposite the double fold into a shape similar to
that of one edge of a plumule leaf. Next, put these leaf
forms together so as to show how the plumule
leaves appear in the bud. Break off the cotyledons and lay the caulicle with its plumule
head up before you so that the plumule turns to
the right. Which leaf seems to overlap the
other?* Are the plumules of the two kinds of
beans alike in this respect? If you can get
beans in the pod find out whether the seeds
growing on one half (valve) of the pod have plumules like
the one figured on this page, while those on the other half
have the overlapping of the leaves reversed. See that your
seeds are kept moist, but not very wet. Put several of
them in a bottle of water at home in order to find out
whether they will grow under water or not.

Fig 8 Com-
mon bean with
one cotyl.don
removed.

Exercise 5.

Examine one of the squash seeds. How many coats are
there? Which one is transparent? cork-like? hard? green?
Why no pocket? Are the cotyledons just alike, or are they
bilaterally symmetrical? Draw two views of the embryo.

---

* The teacher should here have the pupils report by showing of hands
whether their specimens as in the figure on this page, have the lower leaf overlapping.

Taste of it. Crush a bit of it on paper. What substance
does it contain which you did not find in a bean embryo?

Examine a grain of buckwheat (it should be sprouting).
Where does the sprout come out? Find the remains of
some of the flower leaves at the other end. Take the seed
out of the three-cornered pod. Remove the thin seed coat
(why thin?), and very carefully separate the embryo from a
white substance which is packed in the seed coat with it.
Anything found in the seed coat with the embryo is called
albumen or *endosperm.* Taste the endosperm. Feel it with
the tongue and fingers. What common substance do you
think it is? How much endosperm in buckwheat? Partly
straighten the cotyledons and draw the embryo. Put the
embryo by for examination after it has wilted. Look at the
seeds in the jar, and note any changes in their appearance.*

Exercise 6.

Look at the seeds in the jar. What have all the caulicles done? In addition to this, what has the caulicle done

---

* If you are doing this work any time between the first of April and the middle of September you should plant five or six of each kind of seed (the dry ones)
out of doors in good, finely pulverized soil. These should make a row twelve or
fifteen feet long, in which the seeds are about three inches apart. Cover the seeds
with about an inch of fine soil. Separate with sticks the different kinds, and in
some way label them.

If you are studying in the cool months (between September and April), plant
in a window garden of boxes or pots at least one of each kind of seed. Do not
dig them out before they have been up a week or more. Keep the garden well
watered.

Pupils should observe how embryos break through the earth which covers
the seeds, noting the shape of the upper end of the stem, and whether it is the
plumule stem (*epicotyl*) or caulicle (the *hypocotyl* part). [See Ex. 10, p. 17.]

which was pointing upward? If the caulicles of the bean and scarlet runner are an inch or more in length, make dots upon them with a brush and India-ink at equal inter-vals, not much greater than the length of the small letters here used.* This will enable you to tell what part grows fastest. Examine the seeds which you are keeping moist in paper, cloths, sawdust, or sand. What position do the caulicles try to take? Which way do they point? Are rootlets growing from any of the caulicles? If so, from what part? Which seeds have the longest sprouts? Which have none? Try to flatten the cotyledons of your wilted buckwheat embyro. Take also one of those growing in your experimental dish and try to determine the shape of the cotyledons and how they are folded. Do they fit each other? Are they one sided?

By cutting and breaking remove the hard coat of a pine-nut. A membranous coat covers the kernel. What is peculiar about one end? Cut the coat

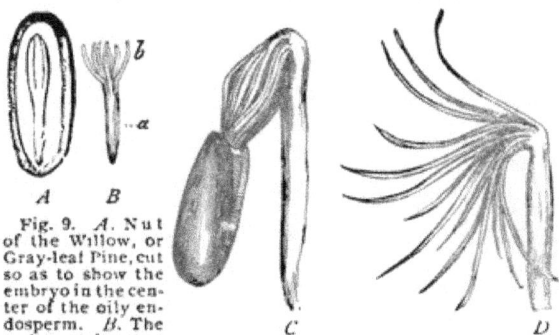

Fig. 9. *A.* Nut of the Willow, or Gray-leaf Pine, cut so as to show the embryo in the center of the oily endosperm. *B.* The embryo taken out and the cotyledons separated. *a.* Caulicle. *b.* Cotyledons. *C.* Germinating nut about to drop the seed-coat from which the cotyledons have absorbed all the endosperm. *D.* The same with seed coat removed.

around lengthwise and split the kernel. Compare what

---

* A pen carefully used will do as well as a brush. The point should not quite touch the sprout, for a slight injury causes it to grow abnormally.

you see with the figure on page 13. The embryo is surrounded by a substance which is evidently not a coat. Which end of the embryo is next to the cap of the kernel? Which is the larger, the embryo or the endosperm? Take the embryo out of its bed of endosperm. Separate and count the cotyledons. Is your pine-nut the same kind as the one figured on page 13? Do the embryo and endosperm have the same taste? Which is the larger? What substance makes a large part of the endosperm? Is there any of this substance in the embryo?

Exercise 7.

Look at the beans in the jar and note their growth.

Study one of the castor-beans as you did the pine-nut. Find the hilum near the white projection *(caruncle)* at one end, and the chalaza at the other. A ridge or thread called

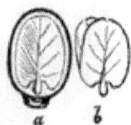

Fig. 10. *a.* Castor-bean cut so as to show the embryo lying in the endosperm. *b.* Embryo.

the *raphe* connects them. With care you can get the embryo out as shown at *b* in the figure. Compare this embryo with that of a sunflower. Taste the embryo and endosperm separately, being careful not to swallow the former. Taste the shell. Which of the three has a pleasant taste? Examine the embryo with a lens and draw the magnified view of it. Why have not the castor-beans and pine-nuts sprouted? What do you know about the uses of these seeds? Have you known animals to be poisoned by eating castor-beans? Look up "castor-bean" in a cyclopedia. Make a list of all the new words you have learned in these exercises.

EXERCISE 8.

What are the beans in your jar doing? Make drawings to show the most interesting facts. Look out for the answers to the following questions from day to day: What part of the caulicle of the common bean grows most? of the scarlet runner? Does the caulicle above the side roots of the bean grow? Does the caulicle between the upper and lower side roots grow? What would be the result if that part grew in a bean which had been planted? Does the caulicle above the side roots in the scarlet runner grow? Which plumule stem grows most? Is the growth of a caulicle up? or down? or both? or neither?

Look at the drawing of a coffee seed on this page. Is the caulicle sticking out of any of your coffee grains? Cut carefully so as to show the small embryo. Describe the endosperm.*

Fig. 11. Grain of coffee cut to show the embryo *a* at one end.

Examine a morning-glory seed. Is there a jelly-like endosperm? How are the cotyledons folded? If possible separate the cotyledons and spread them out. Examine the sprouting squash seeds.

Put ten or more grains each of corn and wheat in water.†

---

* The remainder of the coffee seeds should be planted in a box or pot and kept well watered. Some of them may grow.

† You can make an interesting garden by tying a piece of netting—an old veil—over a goblet or jelly-glass full of water. Place on the netting a few grains of wheat and corn. They will grow and show you clearly what becomes of the endosperm.

EXERCISE 9.

Take one of your grains of soaked corn. Draw the side which has a depression. Divide with a knife into symmetrical halves. Draw the section thus shown. What part is the softest? This is the live part of the seed—the embryo. Carefully remove the embryo of another grain. Try to make out parts. It is evidently very different from the other seeds we have examined. Is the corn plant much like any one of the plants the seeds of which you have examined? Is it more like wheat than like beans or sunflowers?

On the convex side near the smooth end of a grain of wheat find the embryo. Crush the grain with your knife. What is the white powder? Is it part of the endosperm or part of the embryo?*

Draw two views of a grain of wheat twice the natural length, or larger, using a lens to aid you in seeing the small details.

If possible, get an ear of green corn a little older than "just right to roast or boil." The nature of the albumen is obvious in such a specimen, and the embryo is easily removed.

---

* If you have never chewed wheat into " wheat gum," take, this evening, or as soon as you can, a handful of wheat and chew it, while you swallow only what seems to be liquid. (Better put only a few grains at a time in your mouth.) Presently there will be left in your mouth an elastic substance which is made up of *gluten*, colored by particles of the seed coats. The starch was swallowed with saliva.

In the roller process of making flour the embryos of wheat are separated and sold under the appropriate name *germea*.

Compare your drawings with those shown on this page.

Evidently the embryos of corn and wheat are very different from those of the seeds which we have before examined. Instead of two cotyledons there is one enclosing and growing fast to the center of a stem-like organ which is not at once recognized as caulicle and plumule.

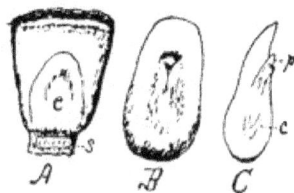

Fig. 12. *A.* A kernel of corn. *s* The stem which usually remains on the cob. *e.* Position of the embryo (chit). *B.* Embryo magnified. *C.* Section of the embryo, showing plumule (*p*) and caulicle (*c*).

Note the work in the jar. Put the remainder of the corn and wheat in sprouting dishes.

EXERCISE 10.

This diagram shows the parts of a *dicotyledonous** embryo which has been growing a week or more. On the right the names and dotted lines indicate the parts manifestly distinct in the embryo before it begins to grow. On the left the growing plumule stem or *epicotyl*, the *hypocotyl*, and the root are shown. The hypocotyl is that part of the caulicle above the side roots. It is the base of the stem, or that portion of it below the cotyledons.

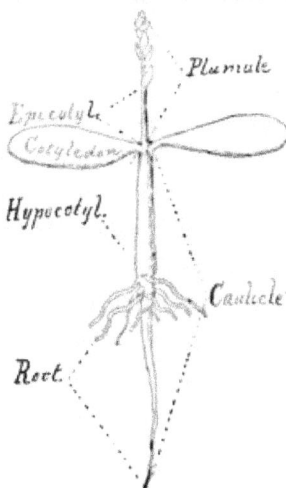

Fig. 13. Diagram showing the parts of a germinating embryo of an **exogen**.

Observe the condition of all

---

* *i. e.*, one with **two** cotyledons. Corn and wheat are *monocotyledonous*.

your unplanted growing seeds. Notice particularly those
which have rootlets growing from the side of the caulicle.
Which of these have developed hypocotyls? In which
have the epicotyls grown most? What changes have taken
place in the cotyledons? Make drawings showing the
present condition of the bean and scarlet runner in your
jar. The India-ink dots should show what portions of the
caulicles have grown most since you marked them. Which
one is like the Windsor bean in its growth? In which has
the epicotyl grown most?

It is important that you find out whether rootlets appear
before or after the hypocotyl begins to grow; and in embryos which do not produce a noticeable hypocotyl you
must note whether the growth of the epicotyl begins before
or after the rootlets start. You must make your seeds tell
you all about the relation between the growth of rootlets,
hypocotyl, and epicotyl, for without this knowledge you
cannot understand how seeds "come up." This first work
of a plant in its second season's growth is not so simple as
is commonly supposed; and to understand how it is done
will try your skill in observing and thinking.*

Exercise 11.

Suppose a bean planted in sand to have grown a sprout
one inch long, as shown in the figure on page 19. Evidently the sand around it must have been pushed out of the

---

* A plant is called an embryo during its first season's growth. This growth
is nourished by the mother plant. [See note, p. 2.]

way as the sprout grew. We have been taught by our growing seeds that the sprout will take a vertical position. We do not yet see what causes it to do so, but we are sure of the fact. Now suppose the sprout grows one inch and is yet erect and smooth, will the tip be an inch deeper in the sand? will the head, bearing the cotyledons, be an inch higher? or will each end move—the one up, the other down —a half inch? This is a simple mechanical problem which you ought to be able to solve. The weight of the bean is not great

Fig. 14. Bean sprouting in sand.

enough to make it sink perceptibly in the sand, so we conclude that gravity practically has nothing to do with the result. The fact that the plant is alive may be important in considering how it keeps upright; but life cannot make a smooth stem move up rather than down, or the reverse, without it does something more than get longer, and keep stiffly erect. If you push up against a ceiling five pounds, do your feet press down five pounds more than your weight? If you place a pin erect upon a potato and push down upon the head with your finger, will the pin go point foremost into the potato or head foremost into your finger? If you reverse the pin and push against the point, what will be the result? Now look at the figure of the growing bean and imagine a real bean in sand or soil. As the stem gets longer will the point push deeper into the sand or will the top with the cotyledons move up through it?

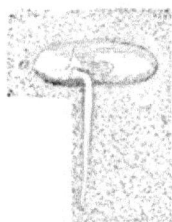

We have learned that before the growing caulicle of a

bean is two inches long a whorl of spreading rootlets appear as shown in the figure on this page.
Let us see how they act when the caulicle
grows. Suppose the hypocotyl lengthens
and the part below the bend remains
straight, what will be the result? Is it
not plain that if the rootlets were removed the result would be quite different? In some of your embryos the
hypocotyl does not grow. How do such
seeds come up? Work upon this problem
of how seeds come up until you have
acquired a thorough understanding of it.*

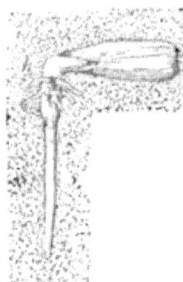

Fig. 15. Sprouting sun-
flower seed, showing side
rootlets below the hypo-
cotyl (4).

EXERCISE 12.

Study your sprouting corn and wheat. If the sprouts
are half an inch long, or longer, taste of the endosperm and
the embryo. What do you think part of the starch has been
changed into? Has this substance been formed in all your
growing seeds? Which sprout appears first, the caulicle
or the plumule? Does more than one sprout come from the
caulicle end of the embryo?

Examine all your seedlings. In which have the cotyledons grown? If the cotyledons do not grow, does the

----

*Since circumnutation of the caulicle only facilitates its movement through the
ground, and has nothing to do in determining which end shall move, it need not
be considered here. The experiments which will enable one to see that all
growing stems circumnutate are not easy enough for beginners. [See Darwin's
"Movements of Plants."]

hypocotyl grow? Look out for exceptions to any general rule which you think you have discovered concerning the relations between the growth of hypocotyl, plumule, and cotyledons.

EXERCISE 13.

Beginning with the first exercise, read again all the preceding pages, and note as you come to them all the questions which you or your seedlings have not satisfactorily answered. Try once more with the aid of new material and new experiments, if necessary, to settle these undecided questions. Perhaps some of your seedlings have not yet had time to show what you wish to find out. Watch and wait; see all that they do, and think why they do one way rather than another.*

EXERCISE 14.

You have now learned that a caulicle has the nature of both stem and root, and that a plumule is a bud which, like other buds, is the beginning of a leafy stem. Your next task is to discover the nature of cotyledons. Seedling buckwheats or morning glories may have already told you. Squash, sunflower, castor-bean, and pine embryos will in time tell the same story; but beans and scarlet runners try

---

* If your work with beans in the jar is unsatisfactory, try again. Scald the jar and the stick by pouring in first a little warm water which you shake, then hotter water, shaking again, and so on till the jar is gradually made hot; then fill with boiling water and let it stand a few minutes. Fix beans on the stick as you did in the first place, and dip them for ten or fifteen seconds in boiling water; then put them in the jar with hot water not quite touching the lowest bean.

to keep the truth a secret. Try to decide this question by a careful study of your seedlings. They must tell you, also, what they do with endosperm when they have any; and how they get along without it when they have none.

Fig. 16. Sprouting wheat on the left; wild oats on the right. *a, b.* Plumule. *c.* The twisted beard which helps to plant the seed.

Make drawings of the present appearance of several of your seedlings. In which have the cotyledons grown most? In which have they not grown at all? What have the cotyledons of the latter done?

Compare your sprouting wheat with the figure on this page. Did the three roots from the caulicle begin to grow at the same time?

EXERCISE 15.

Compare your seedlings; scarlet runner, common bean, and sunflower. The cotyledons of the first do not appear above ground, but those of the other two are pushed up by the hypocotyls after they have securely anchored their bases by putting out rootlets. The cotyledons of the sunflower become broad, green leaves; those of the bean evidently never do any real leaf-work, and, apparently, might as well have remained where the seed was planted.

When do the cotyledons of the bean fall off? Why does not the plumule of the sunflower begin to grow as soon as that of the bean? How are the first two leaves of a bean different from those that appear later? Which of these three plants shows the real nature of cotyledons?

Have your buckwheat embryos yet shown you what they do with the endosperm? Why do they hold on to their coats after they come up?

Draw several of the buckwheat seedlings.

EXERCISE 16.

Compare the roots of your seedlings. Do you find hairs on any of them? Where do the hairs grow? If, in your experiments with beans in the jar, you failed to determine how roots grow, try marking with India-ink the central root of a sunflower after the side roots have begun to grow. Place dots the whole length at intervals of about one-tenth of an inch. Cut thin cross-sections of the hypocotyl and the central root of the largest seedling sunflower. Can you with the help of a lens see any difference in their structure? What do you think can enter the roots? Can particles of earth be taken in by them? What do you think are the uses of roots to the plants?

Fig. 17. Germinating castor-bean, showing the hard coat yet adhering to the inner much-enlarged coat, which forms a sack holding the growing cotyledons and the endosperm, which they are absorbing.

EXERCISE 17.

When your castor-beans come up compare their appearance with the drawing on this page.

Why do the cotyledons remain so long

in the sack-like inner coat of the seed? Your pine-nuts will come up in a similar manner, as shown in Fig. 9. When the shell is thrown off examine the sack within to determine whether the endosperm has been taken up. What absorbs the endosperm?

Fig. 18. A germinating squash seed, showing a "peg" which holds the lower half of the stiff coat while the growth of the bent stem (hypocotyl) above it pries open the seed and liberates the cotyledons.

[*Read and discuss in the class.*]

Germination of Squash Seeds.—

Have you discovered how squash embryos

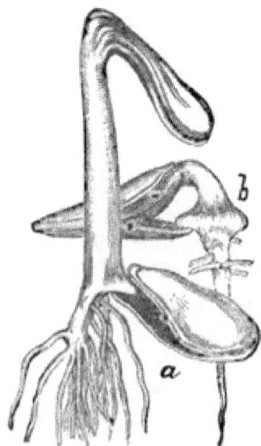

Fig. 19. *a.* A squash embryo just appearing above ground, the elastic coat grasping the "peg" which enabled the growing hypocotyl to pull the cotyledons out. *b.* An embryo which, after sprouting as shown in Fig. 18, was turned over. Then the caulicle again bent downward and grew a new "peg."

pry open their tough coats? Soon after the sprout has gained a foothold in the soil, a little knob grows on the side of the caulicle so as to split more widely open the point of the seed coat, as shown in Fig. 18. Then the hypocotyl between the knob and the cotyledons, by growing, pries the seed still wider open, as seen at *b* in Fig. 19. Finally, by continued growth, the cotyledons are pulled out of the seed coat and upward to the surface of the ground, where they expand, and become pretty good leaves. Seeds planted edgewise, which of course could rarely happen in nature, cannot thus free themselves of their seed coats, and it has been proved by a French botanist (M. Flahault) that seeds which come up with their coats on do not thrive. The seed at *b* in the figure on page 24 was first planted the other side up. It was turned over when the knob on the right had begun to open the seed. The caulicle, which then pointed directly upward, gradually straightened, bent downward, and finally the second knob grew, by the help of which the seed leaves were in a fair way to get out when the drawing was made. Some native California plants get out of their coats in a similar manner.*

Peculiar Germination of the seeds of some plants of the Pacific Slope.—The seeds of Big-root—a pest which grows in nearly every field—behave in a remark-

Fig. 20. Germination of Pig-root. *a.* Planted in a flower pot, the growing petioles forced the cotyledons and seed coat above the surface (*S*) of the soil. *b.* Planted in a shallow box, the growth was lateral. *c.* Natural growth when lying on the ground. *d.* Germinating seed, natural size, pushing the small caulicle (below *f*) into the ground. The plumule is seen in the bottom of the petiole tube at *f.* *e.* Hairs which attach to the soil at the surface.

able manner. The nut-like seeds drop from their prickly pods in June or July, and soon become covered with leaves. The rains of November and December cause them to sprout,

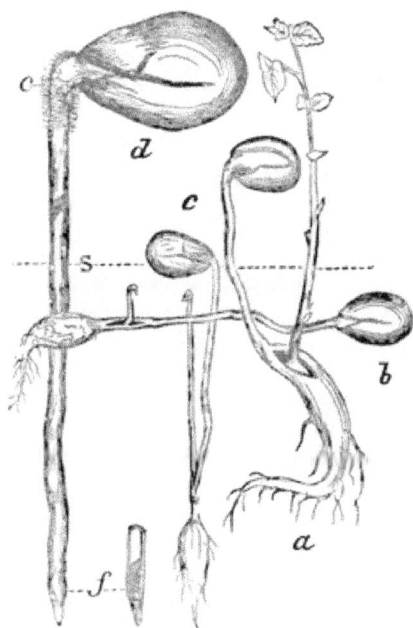

---

* It would be a good experiment to plant ten squash seeds with the flat sides horizontal, just as they would naturally fall, and ten seeds with the sharp ends down. The last would seem the better way, because the caulicles would be pointing downward. The results of such an experiment are very interesting.

as represented at *d*. The mimic caulicle—really a tube
formed by the united stems (petioles) of the thick cotyledons,
and only tipped by the caulicle—penetrates the ground to a
depth, usually, of four or five inches. The plumule meanwhile, as shown in *f*, remains dormant in the bottom of the
tubular sprout. When the petiole growth ceases, the caulicle grows rapidly by absorbing the nourishment stored in
the cotyledons, and becomes thick like a radish. Meanwhile
the plumule begins its upward growth, splitting the petioles
apart, and usually escaping from between them, as shown
in the figure below *c*. In this wonderful way the plumule
bud is deeply planted together with nourishment (stored in
the caulicle), which, if necessary, can be used to aid its
first growth. The reason for this curious behavior is
obvious, when we know that ground squirrels are fond of
these seeds, and that a severe frost will kill the young
plant. If the seeds wait till warm weather to sprout, hun-
gry rodents may find them; if they germinate early, and
in the manner of other seeds, Jack Frost may nip them.*

---

* Dr. Asa Gray, who first experimented with these seeds, found them to grow
as represented at *a*, in the figure (reduced to one-fourth natural dimensions from
Fig. 43, Botanical Text-book, edition of 1879). Evidently on account of some
obstruction, probably the bottom of a small pot, the seeds were elevated two or
three inches above the surface of the soil (the dotted line S represents the surface
of the ground for Figs. *a*, *b*, and *c*). My experiments with seeds planted in shallow
boxes gave very different results—shown at *b*, which is a reduced copy of Fig. 14
of second edition of the Popular California Flora. The plants came up about
four inches from where the seeds were planted, the plumule being pushed later-
ally that distance by the elongation of the cotyledon petioles. Such inexplicable
behavior stimulated to further observation, which resulted in the discovery that
naturally planted seeds, unhampered by boxes or pots, usually grow as represented at *c* and *d*. In one instance a sprout measured seven inches from the
plumule to the cotyledons. The hairs at *e* probably help the sprout to penetrate
the soil by fastening on to the surface crust. Curiously enough, growing sprouts
underground frequently avoid obstacles without touching them. [See Darwin's
Movements of Plants, p. 80.]

Fig. 21. *a.* Tree lupine as it appears germinating in sand. The root hairs are ladened with sand. *b.* Seedling of the small-flowered lupine; the first plumule leaf on the left. *c.* Sprouting seed of the dense-flowered Jupine. *d.* The same after it has been above ground several weeks.

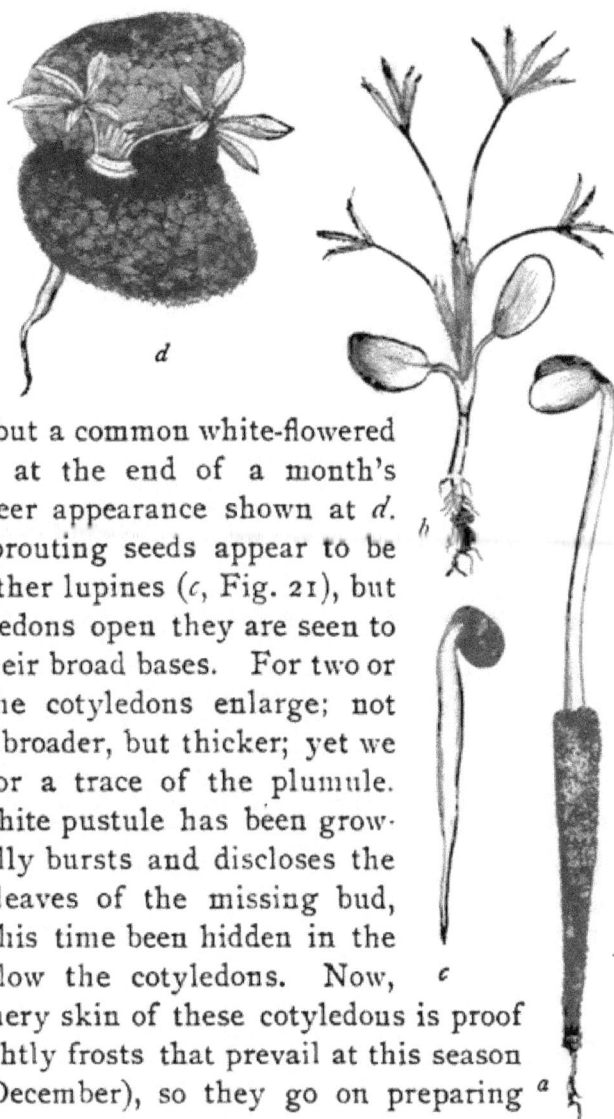

Lupines ordinarily grow as represented in the cut at *a* and *b*, but a common white-flowered kind presents at the end of a month's growth the queer appearance shown at *d*. At first the sprouting seeds appear to be like those of other lupines (*c*, Fig. 21), but when the cotyledons open they are seen to be united by their broad bases. For two or three weeks the cotyledons enlarge; not only becoming broader, but thicker; yet we look in vain for a trace of the plumule. Meanwhile a white pustule has been grow· ing, which finally bursts and discloses the partly grown leaves of the missing bud, which has all this time been hidden in the thick stem below the cotyledons. Now, the tough leathery skin of these cotyledons is proof against the nightly frosts that prevail at this season of the year (December), so they go on preparing food from the air with which to feed the tender plumule, until it also is strong enough to face Jack Frost. If you carefully examine these seeds in various stages of their growth, you will learn that the plumule is at the bottom of a short tube formed by the united petioles of the cotyledons.

EXERCISE 18.

[Read and discuss in the class.]

The seedlings of the Western or Brown's peony dodge Jack Frost in a peculiar way. The small embryo feeds for some time upon its abundant store of endosperm before it bursts the hard coat. The first growth is made by the stems of the cotyledons, which push the caulicle and its terminal bud, the plumule, fairly out of the shell. This movement shows plainly that the embryo does not intend to carry its store of food above ground, as is the custom of castor-beans. Nor does it risk decapitation at the hands of Jack Frost. Neither the hypocotyl nor the epicotyl grows, but a solitary leaf is sent up into the air by the growth of its stem, which, with the undeveloped bud-like blade, looks exactly as we would expect the entire plumule to look. This leaf grows for several

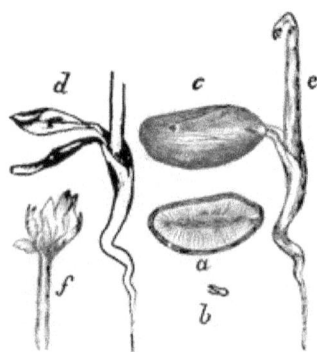

Fig. 22. Germination of Brown's Peony. *a.* Section of the seed showing the minute embryo at the right embedded in the endosperm. *b.* The embryo with the cotyledons separated. *c.* The seed germinating. *d.* The same with the coat removed, showing the growth of the cotyledons *e.* Stem of the first leaf. *f.* The first leaf older and the blade expanded.

weeks before a second one appears, which is followed by others at shorter intervals. Evidently the loss of one or two of the first leaves would not kill the young peony. It is, therefore, able to take advantage of warm weather and moist soil in early spring without the risk of serious injury by late frosts. The leaves prepare starch, which is stored in the roots to help the stem make a rapid growth after all danger from frost is over.

One of our native morning-glories seems to germinate after the fashion of its relatives, but what appears to be the hypocotyl bearing the cotyledons is really the united stems (petioles) of the cotyledons. The plumule between their bases remains inactive for a time and then comes out through a slit in the petiole tube as shown in the figure. This peculiar growth of the petioles and the epicotyl reminds one of Big-root; but in that plant the growing petioles force the caulicle and plumule deeper into the ground, while in the morning-glory the caulicle puts out side-roots which cause the petioles to push the cotyledons up into the air, while the plumule remains nearly where the seed was planted. The scarlet larkspur germinates in a similar manner.

For the next exercise get an entire herbaceous plant of several months' growth. Some weed less

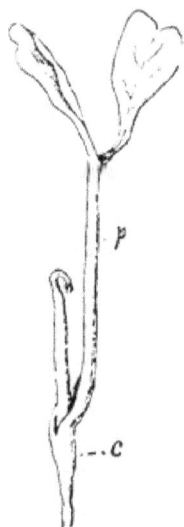

Fig. 23. A native morning-glory in germination. *p.* United petioles. *c.* Caulicle. The plumule not yet above ground. [This is Bentham's *Convolvulus Subacaulis.*]

than two feet high—mallows or filaree,* for example—will best serve your purpose. Do not pull it up, but with a spade or shovel, forced the full length of the blade into the ground, pry or lift the plant out of the ground. Put it with the adhering earth into a bucket of water. Holding it by the base of the stem, move the roots up and down until most of the earth is thus removed, then take the plant out and finish washing the roots in fresh water. It may be left in water over night and taken to school in a newspaper. In washing, do not rub the roots with the hands. Look out for fine hairs on smaller rootlets.

EXERCISE 19.

Study of an Herbaceous Plant.—Observe the roots. Is there any definite order in the arrangement of root branches or rootlets? Look for hairs on the smallest rootlets.†

---

* The two *Erodiums* (*E. cicutarium* and *E. moschatum*) are known everywhere as *filaree*. This name is a corruption of the Spanish *Alfilaria*, but is so well established that to use the Spanish name is pedantic.

† They take up water, which carries dissolved earthy substances to the leaves. Carbon dioxide is absorbed from the air through the upper skin of the leaves. A little of the water, together with the carbon dioxide, is made (digested) into starch in the leaves. Some of the water carries the starch (really the starch is first changed to glucose) to the growing parts of the plant, where it is converted into the various tissues, or it is stored for future use. Most of the water absorbed by roots through the hairs goes out of the leaves through microscopic holes in their lower surfaces. Water and carbon dioxide make more than nine-tenths of all the food consumed by plants. Carbon dioxide is the gas which passes off in bubbles from soda water. When a plant decays or burns it unites with oxygen, forming carbon dioxide, water and a little ashes.

Compare the bark of the roots with that of the stem.
Which is the thicker? tougher? smoother? Is there a
definite arrangement of the leaves? of the branches? What
do you find below each branch? Do you find buds between
the leaves and the stem? The angular space between the
upper side of the leaf or its stem and the stem of the plant
is called the axil of the leaf. Are the branches and buds
axillary? What terminates a branch? Cut a thin cross-
section of the stem near its base and compare it with a section from the upper part. Compare these sections with root
sections. Cut a piece of the stem an inch or more long
and put one end into your ink. Crush a few inches of the
stem and also a part of the root. Which is the toughest in
the center? Bend and double branches and roots. Which
are the more flexible? If the wind blows very hard, what
effect has its force upon the stem? Which roots are tried
by the wind, those on the side next to the wind or those on
the opposite side? Should lateral roots be strong like
masts or like ropes?

Cut thin slices from the stem which you placed in ink.
What part, the skin, the fibers, or the center, absorbed the
ink?

Get for the next exercise a straight sprout from the
roots or base of some tree. Sprouts that have come up
from stumps or from cut roots in an orchard will answer
our purpose. Poplar-trees furnish such sprouts. It mat-
ters not whether the leaves have fallen or not. The sprouts
must not have branches.

EXERCISE 20.

Study of a Woody Stem.—Note the arrangement of the leaves; or, if they have fallen, the leaf scars. Hold the stem so that one of the lower leaves or its scar is on the side next to you. Beginning with the next leaf, count upward to the fifth. On what side do you find it? Count five more. On what side do you stop? Move your pencil point upward from leaf to leaf the nearest way. Is the spiral thus traced like that of a common screw in direction?*

How many turns of the spiral leaf-line from the first leaf to the sixth? Leaves grow from *nodes*, and the spaces between leaves or nodes are called *internodes*. Draw a line directly down the stem from the second leaf to a point opposite the first. In the same way find the point on the first node exactly below the third leaf and so on to the fifth. You have now on one node four points and a leaf or a leaf scar. Suppose the four internodes which separate the five leaves had failed to grow. Do you see that the five leaves would have formed a whorl? How many such whorls would the leaves on the stem make? Do you find buds in the axils of all the leaves? (Above the leaf scars on a leafless stem.) Which buds are largest? What can you say of the size of the terminal bud? Which of the axillary buds were designed to make branches next year? Of what use are the other buds? What is the use of a terminal bud? Draw the terminal bud and the last leaf

---

* If the class have more than one kind of sprout, they should exchange specimens and compare.

or its scar. Carefully remove the scales one at a time until you come to the very tender, real bud in the center, made up of the beginnings of leaves on a stem which has imperceptible internodes. The scales which you have removed are evidently modified leaves, which do work very different from that of normal leaves. We say of these winter-bud scales that they are *homologous* with leaves.*

Cut a thin cross-section of your woody stem. Note in the center the pith, on the outside the bark, and between them the wood. Cut another section near the upper end. What forms the larger part of the old portion of the stem? of the young portion? Does the pith grow any after the stem is a few weeks old? Cut a piece two or three inches long from the base of the stem, and put the larger end into a cup or short bottle containing a little red ink. [The entire class might use one cup or goblet.]

Provide for the next exercise at least one of the short, rough twigs of a cherry-tree. Orchardists call them fruit spurs. [See the next figure.]

The simple microscope and a pair of needles, described in the appendix, will be much needed in the following exercises.

---

* The primary organs of a plant are roots, stems, leaves, and hairs. All other organs or parts are each homologous with one of these four. For example, potatoes, the tendrils of a grapevine, and the spines of a lemon-tree are homologous with stems; the tendrils of a pea, the spines of barberry bushes, and the pods of beans are homologous with leaves; the prickles of rose-bushes, the hooks on cockleburs, and the stings of nettles are homologous with hairs. To find the proofs of these statements is very interesting, but not easy.

Fruit Spurs of Cherry.—Compare your specimen with the figure on this page. Make out the seasonal growths. In what months do cherry-trees grow? When do they blossom, and when is the work of the blossoms completed; *i. e.,* when are cherries ripe? Examine the growth of the last season. How many winter buds were made? Find the scars of the leaves which nourished them. Observe that the terminal bud is smaller than the others. Note the number of visible scales on one of the lateral buds. Carefully remove with a needle all the scales, one at a time. Do not let the point of the needle reach the center at the base where you ought to find three or four tiny buds, which are shown magnified in the figure. If you have a microscope it will be easy to make out the parts of a flower in each bud. Each of the larger brown buds, then, encloses several blossom buds (2 to 5). Do cherries grow two, three, or four together? Dissect the terminal bud. It contains only the rudiments of a stem and leaves. Evidently a similar bud produced last summer the cluster of buds which you are studying. Try to imagine the

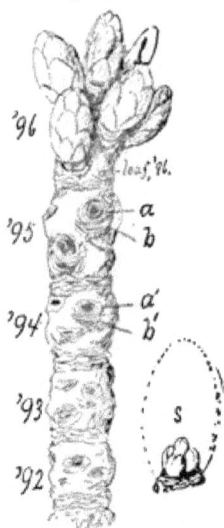

Fig. 24. Fruit spur of a cherry-tree in its winter condition, showing at the upper end the buds which grew during the summer of 1896, with the scars below them of the leaves which fell in October of that year. *a.* Depressed scar left by stem of one of the clusters of cherries which ripened in June, 1896. *b.* Scar of the leaf which fell in October, 1895. *a'* and *b'* show the scars left by cherries of 1895 and the leaf of 1894. *s.* shows the flower buds magnified which are enclosed in the side buds of each seasonal growth.

growth of this terminal bud. What would become of the brown scales? Might not the inner green ones grow a little? Where are the scars left by the scales of last year's terminal bud? Imagine three or four cherries in place of each fruit bud on your specimen. Why not pick the whole compound bunch, spur and all, instead of taking the simple clusters one at a time? Do you see that more than one season is required to make a cherry?*

Gather, before the next exercise, some of the rough, brittle branches of a poplar or cottonwood tree. The seasonal growths on these are often only one or two inches long. The buds are much larger than those on the smooth

---

* If the class is working between January and July, it should continue the study of these fruit-spur buds. When the lateral buds have doubled their dormant size each pupil should examine one. This work would make an exercise in which the following questions, among others, would arise: Which scales have grown? Have they changed in shape? What is peculiar about the margins? the apexes? Have the blossom buds grown? [By this time the class will have studied flowers.] Which organs of the flower are most developed in this bud? Which are longer, the anthers or the filaments? What can you say about the pistil?

The next exercise would be a study of the bud when its recurved scales have disclosed the flower buds, now considerably larger but the peduncles yet very short. These questions would be pertinent: What is the shape of the flower buds? Why are they thus angular? How could you tell by the shape of one bud the number of buds in the lateral winter bud? What changes have taken place in the organs of the flower? Have the peduncles grown? the filaments? Of course drawings should be made. Just before the buds open a third study may be made and these questions asked: How long is the peduncle? Why have the buds lost their angularity? What change has taken place in the stamens? How many ovules are there?

The open blossom, the growing fruit and finally the ripe cherry would furnish material for exercises, completing an interesting series. The pupils would then have a fair knowledge of the seasonal work of a cherry fruit-spur—the highest work of a cherry-tree.

branches. Take home the stem which you placed in red ink, and find out what part absorbs the ink and how high it has risen in the stem.

EXERCISE 22.

Flowering Branches of Poplar.—Compare your specimen with the figure. How is the terminal bud different from the others? What do you think is the use of the balsam on these buds? Draw one of the lateral buds. Remove the scales, one at a time, noting the position of each. Which is the shortest? Which one is narrow and thin? The white cottony centerpiece is a dense mass of blossom buds which when in full bloom form a worm-like hanging cluster of yellowish green flowers. Imagine such clusters two inches in length hanging from each bud.

Fig. 25. Winter buds of Lombardy Poplar (staminate tree). *A.* Flower-bearing branch. *a.* Terminal bud. *b.* Lateral flower bud. *c.* Scar of leaf. *d.* Lateral bud which will produce a branch like the one shown on the right side of the stem below. *e.* A reserve bud. *f.* Ring of scars left by the scales of the terminal bud of 1895. *g.* Scar left by catkin of flowers which fell in March, 1896. *B.* A leaf-bearing branch five years old. *C.* Base and top of a wood-making branch. *l.* Section of a terminal bud. *k.* Section of a blossom bud showing the catkin.

What is the use of the cotton among the buds? Draw the terminal bud. Dissect it. What will such buds on the tree develop into next season? How many leaves were there on the growth of last year? What caused the hollow scars on that and preceding growths? Of what advantage to the tree is it that these branches are brittle?*

What did you learn from the experiment with red ink?

Get branches of buckeye which bear between a pair of twigs a fruit stem (*peduncle*) of the last season, and also a specimen which has a hollow scar between a pair of branches. One specimen may show both conditions.

Exercise 23.

Winter Buds of Buckeye. Find the leaf scars. Are they opposite or alternate? Draw one showing correctly the markings. Buckeyes have compound leaves of five leaflets. A bundle of fibers runs from each leaflet down the stem (petiole) of the leaf. Does the scar indicate the number of leaflets? Before deciding this question examine the leaf scars on many kinds of trees. How many pairs of leaf scars do you find on the stem which bore the fruit? How many on each of the twigs which grew last season? How many buds ordinarily grow on each twig? What, one year ago, was in place of the peduncle and the pair of twigs?

---

* Poplar trees are *diœcious; i. e.*, the staminate and pistillate blossoms are not borne on the same tree. Since the cotton of the seeds is a great nuisance, only the staminate trees are planted for shade trees. It is therefore assumed that the pupils will not be able to get the fruit-bearing branches. When possible, our native cottonwoods, both pistillate and staminate, should be studied.

The numerous dot-like scars on the peduncle mark where little clusters of flowers grew last June. How many clusters were there? How long was the compound bunch of flowers? How many and which flowers produced fruit? Of what use are the flowers which do not produce fruit? Resolve to find out by observing and thinking. Have faith in the usefulness of every part of a plant to the plant itself. Parts of plants or animals which cease to be useful gradually become smaller, and finally, after many generations, disappear. Thus many plants have useless rudimentary organs. These can usually be easily recognized. The kind of organ becoming abortive would be shown by its position; its uselessness by its appearance. For example:— In all the collinsias there are four stamens, each growing on the corolla between two of its five lobes. In the place of a fifth stamen we find a short stem which, bearing no anther, is useless as a stamen and probably has no other use. But in pentstemon the fifth stamen becomes a bar to prevent the entrance of unwelcome insects. In mimulus the place of the fifth stamen is vacant. It has entirely disappeared. Pentstemon furnishes an example of an organ which has taken up a new kind of work.

Dissect one of the terminal buds. Are the scales like the leaves in pairs? How many? Do you find as many pairs of tiny leaves in the bud as there were pairs on the peduncle of last year? Draw the club-like mass of flower buds in the center. Think of the growth

which makes of such a tiny beginning a bunch of flowers
a foot long.*

Secure, if possible, a straight stem three or four years
old. Such stems may be found growing from a stump,
from the base of a poplar, or
in the top of any fruit tree.
Cut it near the base of the
first of the three or the four
years growth.

EXERCISE 24.

**Seasonal Growths of
Stems.**—Cut smoothly the
base of the stem. You thus

Fig. 26. *a.* Buckeye, winter bud.
b, One of the inner pair of scales
which has, in its spring grow h.
developed leaflets, showing that
the bud scales are petioles.

expose to view a section of the oldest as well as the newest
wood. This section should, if you have followed directions,
be near the base of the oldest growth. If each seasonal
growth, after the first, forms a layer of wood outside of

---

* Sometimes only one of the pair of buds at the base of the peduncle grow;
and rarely neither of them grow the season after their formation. When they
make but a short growth the terminal buds may not produce blossoms.

The spring growth (in February, March, and April) of winter buds should
be studied by classes working from January to June. None of our native trees
are more interesting than the California buckeye. The bud scales grow considerably, as shown in the figures here explained, and thus furnish proof of the
fact that they are of the nature of leaves, but represent only a part of the leaf.

The winter buds of sycamore are very interesting, and should be studied if
material can be obtained. The scales form a series of complete caps. Before
the leaves fall the buds are completely hidden by the petioles which enwrap them.

When it is difficult to obtain specimens from native trees such cultivated trees
as maple, box-elder, tulip-tree (very interesting), and locust may be studied.
Lilacs have large buds.

that formed the preceding season, the nearly tubular growths would appear in a cross-section as rings. You have estimated the age of your specimen by the bands of scars produced by the scales of terminal winter buds. If your specimen bears two such bands, you have decided that the base contains pith and a layer of wood made three years before its outer layer was formed. Do the rings of the section confirm your decision? Make a section half an inch below the base of the next growth and compare with one made just above the base. Why do they show the same number of rings? To understand this you must imagine the condition at the end of the first season's growth. There was a terminal bud where now you find the band of scars which marks the end of the first growth and the

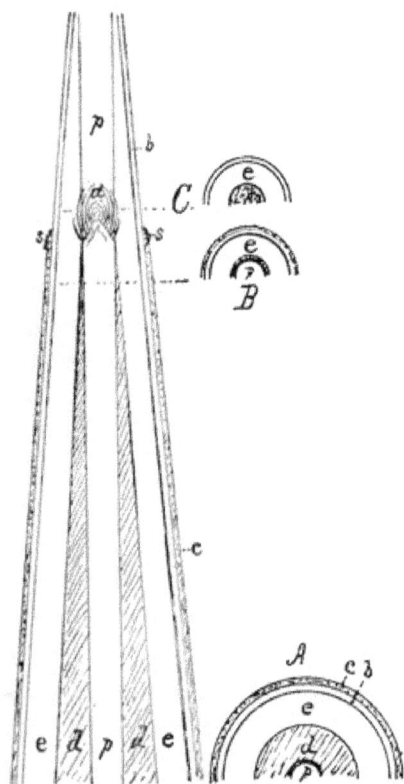

Fig. 27. **Diagram** showing the seasonal growth of an exogen. *a.* Bud at the end of the first season's growth, the wood and bark of which is shaded. *b.* Bark of the second season's growth enclosed below the scars (*s*) of the first season's bud scales by the bark of the first season (*c*). *d.* Wood of the first season. *e.* Wood of the second season. *A.* Cross-section near the base. *B.* Cross-section near the end of the first season's growth. *C.* Cross-section just at the beginning of the second season's growth (the pith wanting).

beginning of the second. Do you see that the layer of wood just below the bud would be very thin, and, therefore, not easily seen?*

In the figure on page 40 the section at A shows two seasonal growths, but at B, near the end of the first growth, only one—that of second season—can be easily made out. Is the pith as large in the growth of the last season as in that of the first? What change must take place in the bark as the stem increases in thickness? Do you see any signs of its stretching? Can it stretch enough to cover the stem when it has grown to be two feet in diameter? Is the bark any thicker on the trunk of a tree than it is on the twigs? If it grows in seasonal layers, where is the youngest layer? What becomes of the oldest layer? Sequoias live to be over two thousand years old. Is any of the bark on a Sequoia that old? Why do you think not?

Examine, this evening, bark on large trees or sticks or on logs of wood. Count the layers of growth of the wood, and thus estimate the age. Look at the end of a board and

---

* In the diagram on page 40, the thickness of growth is represented as very much greater in proportion to the length than is ever the case. It shows more plainly than a real section that an ideal exogenous tree has a trunk made up of a succession of elongated hollow cones, one over the other; the oldest one forming the center for only a few feet at the base; the youngest enveloping all the others from base to summit. This condition is nearly realized in coniferous trees. A pile of hollow cones, all the same size, represents roughly the condition of a palm-tree in which the trunk, made up on the outside of the bases of the equal cones, is as large at the top as at the bottom—or larger, if the younger growths were better nourished. Only the upper (youngest) cone bears leaves, and these send fibers down between the fibers of older growth for some distance.

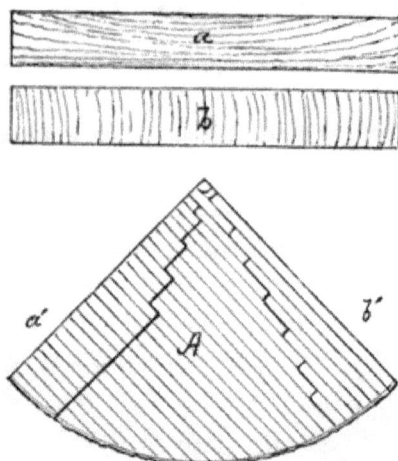

Fig. 28. *a*. End of a board cut nearly parallel with layers of growth. *b*. End of a board cut nearly parallel with the medullary rays. *A*. End of one-quarter of a log showing the character of the boards sawn from it. The straight lines show the saw cuts and the curved lines represent direction of annual growths. Boards like *a* would be cut on the left, and boards like *b* on the right. [See note in Appendix.]

determine where the center of the log must have been from which it was sawn.*

EXERCISE 25.

[Read and discuss in the class.]

Forms and Habits of Stems.—Plants which die, at least to the ground, after they have matured fruit are usually called *herbs*, and their stems are said to be *herbaceous*.† Many such plants live but one season (usually five or six months) and are called *annuals*.

* Pupils may, by trial at home, determine for different kinds of wood whether they split more readily along the lines of growth, along lines radiating from the center, or along lines making an angle of forty-five degrees with these. A piece of flooring, showing annual growths at the end, as seen in the figure at *a*, and another, with the growths as shown at *b*, might, with the well-worn floor of the schoolroom, enable the pupils, and the teachers as well, to learn a practical lesson on "How to select boards for the floor of an uncarpeted room." Find which boards have worn best in the schoolroom; those with the "grain" like *a* or those like *b*. It is plainly shown in the figure that only part of a log sawn in the usual way can be made into good flooring.

Something about driving nails may be learned too. Evidently a nail driven perpendicularly through boards, like *a* or *b*, would be more likely to split them than if driven obliquely; but in a board with either the radial or the growth lines not parallel with the sides, a nail should be driven perpendicularly.

† Half woody plants, like blackberries, often produce a stem one season which after fruiting the next season dies. Such plants are not called herbs.

Some herbs in the temperate zones lay up a store of food in their stems, leaves, or roots the first season of their growth from the seed. This is consumed the next season in the work of producing fruit. Then the plant dies. Such herbs are *biennials.* Many plants on the Pacific Coast, like radish and lettuce, store food in the fore part of the season to be used in the latter part. When a plant comes up year after year from some portion that lives underground at least from one growing season to the next, it is called a *perennial* herb. Some plants which are annual herbs in the Atlantic States live several, or even many years on the Pacific Coast, becoming semi-woody or tree-like perennials. Castor-bean and tomato are notable examples.* Plants with woody stems if small and prone to grow in clumps of many stems from one root system are called *shrubs.* When a stem or trunk stands apart from others and in time attains a diameter of several inches or a height of twenty feet or more it is called a *tree.* The same kind of plant may be a shrub in one locality and a tree in another. Shrubs are usually known as bushes. When growing close together over considerable areas they form *chaparral* or thickets. Trees form *groves* and *forests. Deciduous* shrubs and trees shed their leaves at the close of the growing season and remain leafless during the resting season. *Evergreens* retain the old leaves until new leaves are formed.†

---

* Century plants. like annuals and biennials, die as soon as the fruit ripens; but numerous *suckers* take the place of the old plant.

† Cone bearers may retain their leaves for several, or, as in the araucarias, for many years.

Weak stems, whether herbaceous or woody, may creep,
trail, run, twine, or climb. Bermuda grass and other plants
rooting at the joints are *creepers.* A strawberry has *runners.* Morning-glories and some beans are *twiners.* Get
roots like beets or radishes for the next work.

EXERCISE 26.

Roots and Their Work.—Roots grow out of stems.
Seedlings, as you have learned, have roots which grow from
the caulicle or stem of the embryo. At first there is a
single root which is apparently but a continuation of the
stem. From this grow secondary roots. If the central
root continues to be much larger
than the others it is called a *tap
root* (*a* in Fig. 29). Many plants
have *fibrous roots,* as shown at *b*
in Fig. 29 and on p. 50, 51. Plants
of the lily family, the cultivated
grains, and all grasses have fibrous
roots. Often starch and sugar are
stored in roots instead of in bulbs,
etc. When tap roots are thus distended they become *spindle-shaped,*

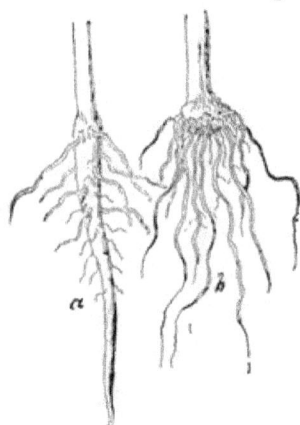

Fig. 29. Roots. *a.* Tap root. *b.*
Fibrous roots.

as are carrots, or *napiform,* which
is the shape of a turnip. Secondary roots, like those of
sweet-potato and dahlia, may be in clusters or bunches
called *fascicles.* The underground system of roots, rootlets, and root hairs corresponds to the stems, branches,

and leaves above ground. As leaves usually live but one season and grow only on stems of that season's growth, so root hairs grow only on young rootlets and last but for a season. They take up water with dissolved mineral matter which goes to the leaves. There it meets carbon-dioxide, which the leaves absorb, and the two are digested or made into starch from which, after being changed into sap, the plant tissues are made.

Make a drawing of your root, showing, if possible, the bases of the leaves which grow from it. Where are the rootlets? Make cross and vertical sections. Draw them. Are there any lines of growth? What do you think caused those lines? What is the nature of the material stored in the root? What normally becomes of this material? In what condition would you expect to find the root of a beet or a radish after it had gone to seed? Why do wild radishes bear more seed than those which are cultivated? Do these plants—turnips, beets, etc.—have as large roots in the wild state as when cultivated? How does a gardener set about to improve the size, flavor, etc. of any of these roots?

EXERCISE 27.

[Read and discuss in the class.]

The Secondary Organs of Plants.—You have learned that the primary organs of a flowering plant are stems, leaves, roots, and hairs (see note p. 33). You have also had some evidence of the fact that all the other organs of a plant are modified forms of these four. This fact should be kept constantly in view by the student. It is the key which opens the way to a clear understanding of all that is most interesting and wonderful in the structure and behavior of plants.

Whenever it becomes necessary for the good of a plant that something should be done different from the normal work of its primary organs one of these fits itself, as it were, for the new office. Some time in the course of the life of any seed-producing plant one or more of its buds develops into a flower or a cluster of flowers, the parts of which are modified stem and leaves. In a few plants some of the buds become tendrils, the undeveloped leaves remaining useless bracts, or disappearing. In others the leaves, or parts of them, form tendrils (Fig. 48), or, remaining nearly normal, as in nasturtium, clematis, and the jasmine-like solanum, do the work of tendrils. When it is necessary to store up food for the renewal of life after a season of leafless rest, or to carry the plant through a period of drought, stems or leaves become storage organs. In some plants branches burrow in the ground and become tubers (Fig. 33). In others, the base of the stem, consisting of many short nodes, becomes a fleshy rootstock or corm, or, developing only the leaves of the base, a bulb is formed (Fig. 36). Besides these there are many less common but often very useful organs. Bring for the next exercise a potato.

Exercise 28.

Storage Stems.—*Study of a Potato.*—Find the stem end of the potato. The other end is the apex. Can you make

out any order in the arrangement of the eyes? Where are
they most numerous? What do you think the eyes are?
Do you know what they will do if the potato is planted?
Do they ever grow if it is not planted? If stems grow from
the eyes, the eyes are buds. Normal buds should have leaves
below them. Find the rudiment of a leaf—the eyebrow.
Buds are the outgrowth of stems. Potatoes, then, are
homologous with stems. Such underground stems are
called *tubers.* Cut the potato in two crosswise through the
center of an eye. Put one piece, cut side down, on
a plate, and pour around it a spoon-
ful of ink. (Red ink if you have it.
Several pupils can use one plate.)
Cut a very thin slice from the other
piece and hold it up to the light.
Draw the section. Make a section
through another eye. Does the line
in this also run to the eye? We

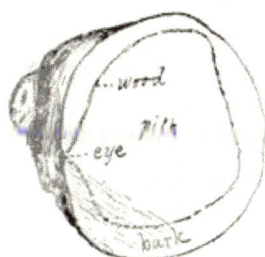

Fig. 30. Section of a potato.

have learned that stems are made up of bark, wood, and
pith. In our experiments with ink only the wood absorbed
ink readily. What do you think is inside the line? What
between it and the skin? Now, see what the piece in the
ink says about it. First cut off a very thin slice, then
another, which you examine. What part of a potato has
wood fibers? Did you ever see this thin layer of wood
conspicuous in a boiled potato? Evidently only the outer
layer of bark is the skin. A potato, then, is a stem almost
entirely made up of pith and inner bark. These parts are

made up of little sacs called cells which are filled with wet starch. This starch was made in the leaves of the potato plant and stored in its tubers for the use of the buds in their first growth the next season.*

The use of starch stored in tubers, then, is similar to that of starch stored in cotyledons; or, as endosperm, within the seed coat outside the embryo. It feeds the plant in its growth from a practically leafless condition (as a bud on a tuber or an embryo in a seed) until it has leaves with which to make its own starch. Starch is frequently stored in stems above ground. Sago, a kind of starch, is obtained from the pith of the sago palm. Starch and sugar are stored in such roots as sweet-potatoes, yams, beets, etc.

---

* The teacher might profitably have ready for this exercise a piece of clean muslin two feet square, a large tin grater, two tin pans, a bucket and something with which several pounds can be accurately weighed.

Take two pounds of the larger pieces of potato (used by the pupils but not inked), and grate about half of each piece into the pan which should be half full of water. Continue grating until just a pound is left. Stir the grated potato until it is mixed with the water. Place the cloth over the empty pan and pour the grated potato and water into the cloth-lined pan; bring up the corners of the cloth and hold them with the larger part of the cloth in the left hand while you knead under water the starch bag thus formed until you think all the starch is out. Take the bag out and squeeze it over the pan until it does not drip. Then, if you wish to be accurate, put water in the first pan and try to get out more starch by thorough but not too energetic kneading. When the starch has settled so as to leave the water clear, pour off the water and set the pan on edge to drain, and leave it until the starch is dry (some time the next day). Put the celulose which was left in the cloth on a plate to dry. The dry starch and the celulose should be weighed. The sum of their weights taken from one pound shows the amount of water in a pound of potato. This work would make a separate exercise.

EXERCISE 29.

[Read and discuss in the class.]

Storage Stems and Leaves.—
Perennial herbs have persistent underground stems from which aerial stems,
or at least leaves and flowers, grow
every year. These stems may usually
be distinguished from roots by buds or
by rudimentary leaves. They all contain more or less starch, particularly at
the close of the growing season. Those

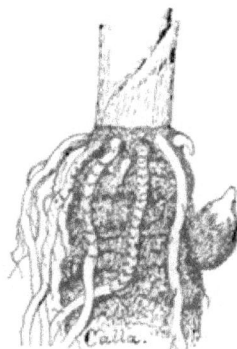

Fig. 31. Vertical rootstock of calla.

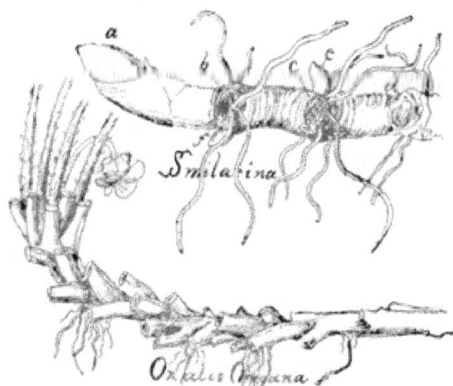

Fig. 32. Horizontal rootstocks of smilacina (false
Solomon's Seal) and Oregon oxalis. *a.* Bud for the
next growth of leafy stem. *b.* The scar left by the
above-ground stem of last year. *c* and *d.* Similar
scars of preceding years. *e* and *f.* Dormant buds.

which are rootlike, or
somewhat woody, or not
much distended with
stored food are usually
called *rootstocks** (See
Figs. 31 and 32). In the
last exercise we learned
that stems like potatoes,
and those shown in Fig.
33, are called tubers.

Tubers, unlike most
fleshy rootstocks, are a
means of multiplying the

---

*Our most troublesome weeds such as field morning-glory, known in
England as bindweed, prostrate heliotrope, or blueweed, and licorice, common
on rich bottom land, are not easily exterminated because their numerous long and
slender rootstocks are not injured but benefited by ordinary cultivation which
makes cuttings of them and thus multiplies and spreads the plants.

plant. The smilacina rootstock produces usually but one bud to succeed the old plant; but a single potato plant produces many tubers, each with several eyes.

Corms are short vertical rootstocks similar to tubers, but each node produces true leaves which appear above ground, and only a single strong bud produces a new plant. The bases of the leaves form several thin coats. Fig.

Fig. 33. Tubers of Scutellaria tuberosa. *a.* The old tuber which has given up its starch to feed the above-ground stem in its first growth. *b.* An underground stem just beginning to thicken and thus form a tuber. *c.* Tuber terminating a stem which grows from the axil of a bract on the upward growing stem just below the surface of the ground. Roots also grow from this stem. *d.* A pair of bracts. *e.* A three-jointed tuber.

34 and Fig. 35 show corms which resemble bulbs and are usually called solid bulbs.

Bulbs resemble winter buds, but are much larger, and instead of thin dry scales the juicy thickened bases of the last season's leaves surround the real live bud. If these bulb scales are broad, so that each envelops all within, a *tunicated bulb* is formed. Narrow and very thick scales make a *scaly bulb.*

Fig. 34. Half-solid bulb of Brodiæa capitata with young bulbs which terminate short rootstocks.

Common cultivated onions have tunicated bulbs. Most lilies have scaly bulbs. Some of our Fritillarias have, besides a few large scales, numerous small

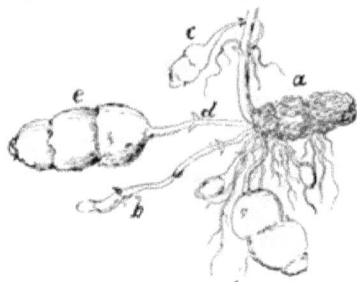

Fig. 35. Half-solid bulb of Allium serratum.

ones which resemble grains of rice.* (See Fig. 38.)

EXERCISE 30.

The Forms of Leaves.

—Leaves are very important organs, since it is their office to change plant food (chiefly carbon-dioxide and water) into substances (chiefly starch) which after further changes are used to build up the plant. This work is like that of the stomach in animals. Leaves also do a kind of work for the plant similar to that done by the lungs of animals.

Guided by Fig. 40 arrange your leaves in three sets: (1)

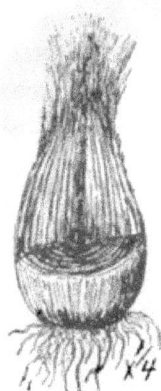

Fig. 36. Tunicated bulb of soap root.

Fig. 37. Scaly bulb of a lily.

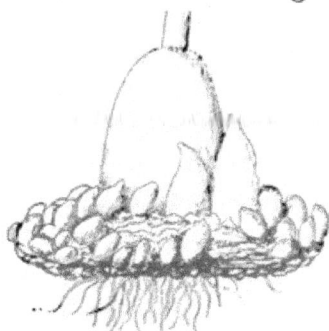

Fig. 38. Bulb of Fritillaria lanceolata.

* The development of rootstocks and bulbs should be studied by pupils whenever opportunity offers. They might find out what becomes of the old bulb and how new ones are formed, whether a single scale of a bulb can be made to grow or not, etc. Soap root has a very interesting bulb and should be obtained if possible. Cut crosswise, its halves make very effective flesh brushes and at the same time furnish a better cleansing material than soap. The outer coats of many native alliums are interesting objects under a microscope. The rootstocks of the European or field morning-glory, now a common pest in the coast valleys, would be good material for an exercise. Pupils might tell what they know about the habits of the plant and the methods used to extirpate it. The teacher might try the iodine test for starch upon any of these storage stems.

Putting those together which have the three parts there shown—*blade, petiole, stipules.* (2) Those which have only blade and petiole. (3) Those consisting of blade only. The first are said to be *stipulate* and *petiolate*, or *petioled;* the second are peti-

Allium unifolium, Kell.

Fig. 39.   A peculiar onion.

olate, and the third are *sessile.* Note whether the stipules are leaf-like, or thin and dry *(scarious),* or rudimentary.

Now rearrange your leaves in two sets: (1) those broadest in the middle; (2) those broadest near one end. Compare the first, one at a time, with the outlines in Fig. 41. Compare each of

Fig. 40.   Creeping wood violet.

the second with the form shown in Fig. 42. Then match each leaf, if possible, with a form in Fig. 43 or in Fig. 44, laying aside those which you cannot match. Next, name the forms of your "misfits," with the aid of the following:

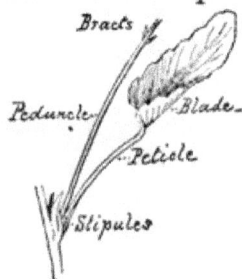

Fig. 41.

A form midway between linear and oblong is *linear-oblong.* If nearer oblong it is *narrowly oblong*; and if nearer linear it is *broadly linear.* So, too, we may have *broadly oblong,* or *elliptic-*

oblong, and *broadly elliptical.* In a similar way we name intermediate forms of those shown in Fig. 42. The meaning of *lance-ovate*, *narrowly lanceolate*, *broadly ovate*, etc., is obvious.* If you have a leaf lanceolate in outline with the stem at the narrow end, it is *oblanceolate*, as shown in Fig. 44. You may possibly have the forms *obovate*, which is the reverse of ovate. Unusual but possible forms are *obreniform*, *obcordate*, and *obfalcate.*

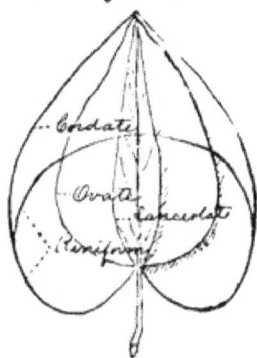

Fig. 42.

Save the leaves you have just studied and bring for the next exercise as many other forms as possible. Get also as many types of margin as you can (see Fig. 45).

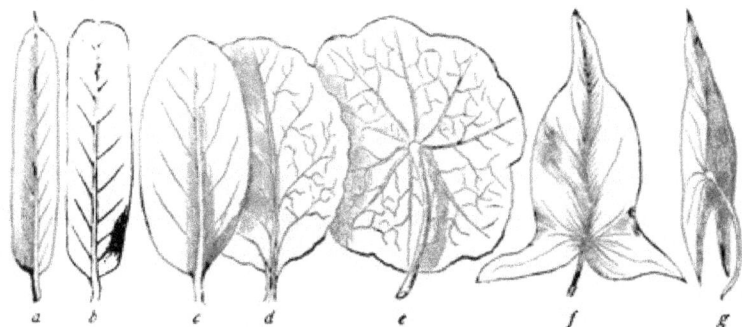

Fig. 43. *a.* Linear. *b.* Broadly linear. *c.* Elliptical. *d.* Broadly elliptical. *e.* Peltate. *f.* Hastate. *g.* Sagittate.

---

* It will be a good exercise to cut these forms out of paper. First, cut into the orbicular form, and then, by trimming the sides, make in succession the forms: broadly elliptical, elliptical, elliptic-oblong, oblong, narrowly oblong, broadly linear, linear, narrowly linear. In a similar way, beginning with ovate, cut out the base to cordate, cut off the upper part to reniform, etc.

EXERCISE 31.

The Forms of Leaf-Margins, Apexes, and Bases.— Again classify all your leaves according to form. Such forms as *triangular, wedge-shaped (cuneate), sword-shaped, fiddle-shaped, fan-shaped* do not need to be defined here.

Fig. 44. *a.* Lanceolate. *b.* Oblanceolate. *c.* Spatulate. *d.* Ovate. *e.* Cordate. *f.* Reniform. *g.* Cimeter-shaped. *h.* Halberd-shaped. *i.* Eared at base.

With the help of Fig. 45 name the margins of your leaves. Observe that a serrate margin becomes crenate if the teeth are rounded. If they are but slightly rounded then call the margin *crenate-serrate*. Margins may have large teeth which are themselves toothed. Thus we have *doubly serrate, doubly dentate,* etc. The words coarsely and finely are used to express conditions varying from the type. Note the apexes. When about as sharp as in the leaf at *h*, Fig. 44, they are *acute*. In the same Fig. *a, e,* and *g* are *acuminate, c* and *d* are *obtuse*. When the end seems as if it were cut off square it is *truncate*. A shallow notch in the end of an obtuse leaf makes it *emarginate*. A *cuspidate* leaf ends in a short rigid point (*a*, Fig. 43). A mere prolongation of the midrib into a sharp point makes the leaf *mucronate*. The bases of leaves may be acute, acuminate, obtuse, or truncate. Arrange your classified leaves in order on sheets of paper, writing under each a brief description.

Are any of the leaves one-sided? Did any of them grow in such a way that one edge was nearer than the other to the stem on which they grew? Are the petioles of any of the leaves flat? Which way are they flattened, at right angles with the blade, or so as to be parallel with it? Did any of the leaves hang nearly vertically?

Get for the next exercise, if possible, leaves of elm, poplar, eucalyptus, pepper-tree, buckeye, geranium, ivy,

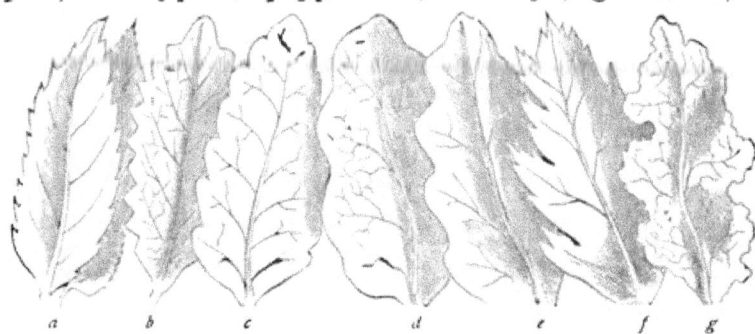

Fig. 45. *a*. Serrate. *b*. Dentate. *c*. Crenate. *d*. Repand, or undulate. *e*. Sinuate. *f*. Incised. *g*. Erose.

clover, and any forms not already procured. Note particularly their position on the stem.

EXERCISE 32.

Lobed and Compound Leaves.—Classify all your leaves putting them into three sets: (1) those which are entire; (2) lobed leaves (see *a*, *b*, *c*, and *e* in Fig. 46);

Fig. 46. *a.* Pinnately lobed. *b.* Pinnately parted. *c.* Palmately lobed. *d.* Pinnate. *e.* Pedate or palmately parted.

(3) compound leaves (*d* in Fig. 46 and Fig. 47). Classify
the first set as you have done before. Classify the second,
putting those with leaflets all at the end of the petiole (see
Fig. 47, *a*) separate from those like *d* in Fig. 46.

Have you observed anything which will enable you to
answer these questions? Which edge of an elm leaf is
nearest the twig on which it grows? Which are moved
most by a breeze, elm leaves or poplar leaves? Why?
Which have stipules?
Stipules which soon fall
are *caducous*. Describe
the lobed leaves. Which
have acute or acuminate
lobes? Rounded or ob-
tuse lobes? Observe the
spaces between lobes
(sinuses). Which are
rounded? Which are
acute? Notice the veins
or ribs which run from
the base of some lobed
leaves through the lobes.

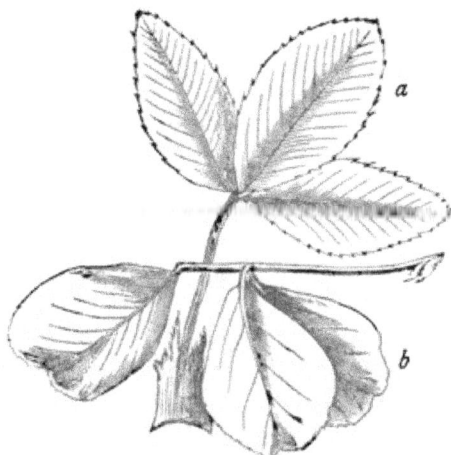

Fig. 47. *a*. Palmately 3-foliolate leaf of clover
with lacerate adnate stipules. *b* Pinnately 3-folio-
late leaf of bur-clover with free lanceolate stipules.

(Fig. 46 *c* and *e*.) These leaves are *palmately* lobed.
Those, like *a* and *b* in the figure, are *pinnately* lobed. When
in any leaf there are three or more large veins or ribs,
which starting from the base of the blade separate widely,
we say it is palmately veined. If the large veins are
branches of a midrib the leaf is pinnately veined. Deeply

lobed leaves are pinnately or palmately *parted* (Fig. 46, *b*
and *e*). *Divided* leaves are cut quite to the midrib if
pinnately, or to the base if palmately divided. Leaves are
cleft when the sinuses between the lobes are sharp. It is
common to give the number of lobes in the descriptive
phrase. In Fig 46, *c* is palmately five-lobed, and *e* is
palmately five-parted. Pinnately lobed, parted, and divided
leaves come under the general term *pinnatifid.* When the
lobes are pinnatifid the leaf is *bi-pinnatifid.*

Compound leaves have distinctly separated *leaflets,* which
are joined to a common petiole in *palmate* leaves (Fig. 47,
a), and to a prolongation of it called a *rachis* in pinnate
leaves (Fig. 46, *d*). Clover leaves are usually palmately
3-foliolate. Bur-clover is pinnately 3-foliolate (Fig. 47).
An *abruptly* pinnate leaf has no terminal leaflet. Leaves
may be twice or thrice compound; that is, the leaflets may
be compound. Such leaves are said to be *decompound.* In
a *tri-pinnate* leaf the leaflets of the leaflets are pinnate.

Keep all your leaves * and find if you can other kinds
for the next exercise. Leaves of locust, ailantus (tree of
heaven), acacia, lupine, strawberry, elderberry, etc. may
be added to your collection.

EXERCISE 33.

Decompound Leaves.—Have you a compound leaf,
with compound leaflets; that is, decompound ? Is it of the

---

* Do not be so slovenly as to put leaves or flowers in any useful book. Find
one useless for any other purpose; an old almanac, for example. A blank book
in which finally to fasten the leaves after they are dry is desirable as a record of
your work.

palmate or pinnate type? Most decompound leaves of the
former type divide continually into three parts forming a
ternately decompound leaf. Columbine and thalictrum
(meadow-rue) have such leaves. Some acacias have pinnately decompound leaves. Have all the leaflets petioles in
such leaves?

Study the leaves which you have not been able to
describe; that is, try again. Have you a palmately decompound leaf with pinnate leaflets? A leaf pinnate below
and pinnately lobed above? A leaf with a few leaflets at
the base of an entire blade? A leaf with three or more
ribs running from base to apex? Try to make out the
plan of the skeleton first. Then tell how the leaf is built
upon the skeleton. Compare forms to those of common
objects. Make descriptions as brief as possible, using the
adjectives with which you have long been familar, as well
as those which you have learned in these exercises. Usually the most important characters should be first given,
but you must pay some attention to sound, as well as sense.
Write out a description of an elderberry leaf (or any pinnate leaf), and then read the following: Elderberry leaves
a:e pinnately 7-foliolate, extipulate and borne on petioles
shorter than the spaces between the pairs of thin, smooth
leaflets, which are nearly elliptical, and serrate from near
the acute base to the abruptly acuminate apex. The lateral
leaflets are sessile, or nearly so, and slightly one-sided, the
upper edge being shorter than the lower. The general outline of the entire leaf is ovate, the length on old trees eight

or ten inches, and that of the nearly equal leaflets about three inches.*   Bring for the next exercise grape-vines.

Exercise 34.

Grape Tendrils.—Examine carefully your specimen of a grape-vine. Remember, branches normally grow from axillary buds, and that, therefore, we expect to find below a young branch a leaf. Note that the tendrils are not in the axils of the leaves, but opposite to them, that the branches of these tendrils are subtend by bracts, which, you know, are rudimentary leaves, and that the upper tendrils are longer and evidently older than the stems from which they seem to spring. We must conclude that, since only stems bear leaves or bracts, grape tendrils are stems. Moreover, not being in the axils of leaves, they must be primary stems. Producing but one or two leaves a grape stem develops the succeeding internodes into a branching tendril. The branch at the last normal node becomes the leading stem which at the first or second node becoming a tendril, gives place in turn to its branch, and so on to the end of the vine. In your older and longer specimens there are near the base clusters of flowers or grapes. Their posi-

---

* The work on leaves may profitably be continued for several more exercises. Out of school, pupils should observe the habits of leaves. Many leaves change their position or the relative positions of their parts at night. Oxalis leaves and those of most acacias change remarkably. The common mallows leaf when young faces the sun all day. Because of the hanging habit of eucalyptus leaves, aided by their smooth surfaces and acuminate points, they probably send more of a passing fog to the ground than any other tree, besides letting all rain drops slip easily by.

tion is exactly that of tendrils. Some of their basal branches are tendrils, and others partake of the nature of both tendril and flower stem. Evidently the tendrils and flower clusters of grapes are homologous with the

Fig. 48. *A*. Apex of a growing grape-vine. *a*. The main stem developed as a tendril. *b b*. Bracts in the axils of which grow branches of the tendril. Above is another tendril surpassing the branch which seems to be the main stem. *c*. Reserve or winter bud. *d*. A bud which becomes a short stem bearing a few leaves. *B*. A grape tendril which was called upon to support a heavy vine. *C*. The pinnate leaf of a sweet-pea, showing all but one pair of leaflets developed as tendrils.

primary stems. Probably the ancestors of grapes had no tendrils but in climbing used the flower clusters as holdfasts.

Observe the buds. Since all the normal axillary buds develop immediately, we would expect to find in your speci-

mens only one bud, and that at the last node in the axil of
a leaf not yet unfolded. We do find, however, two buds in
the axil of each of the upper leaves; and lower down, perhaps, one of these has developed into a short branch bearing
one or more leaves. There are then in the axil of a very
young leaf, at the tip of each growing vine, three buds. One
immediately develops into a continuation of the vine whose
stem there changes into a tendril, another soon becomes a
short leaf-bearer, and the third waits until next season when
it may or may not be called upon to do the work of vine-
making. In vineyards all the buds left after pruning
usually grow. Observe, if possible, what becomes of the
tendrils which do not help to hold the vine. Fig. 48
represents a tendril which had much to do.*

Exercise 35.

Flowers and Their Organs.—Plants must die, therefore they provide for the continued existence of their kind.
Since from the beginning of plant life this has been a necessary work, the organs for accomplishing it have reached a
higher condition than those used for any other purpose.
The highest work done by the plant, therefore, is the making of seed, and the highest plants are those which have the
seed-making organs—the floral leaves—most changed from
their original forms in the way that will enable them to

---

* The tendrils of peas, squash, melons, passion-flower, Virginia creeper and
other plants may be studied out of school and the results placed in the notebooks.
Some are homologous with branches and some with leaves or parts of leaves.

make with the greatest economy * the best seeds, provided
with the best means of transportation, at the least expense
to the fields most fitted to their needs.

Perhaps you have already learned that flowers produce
seeds. You have observed that something in the center of
a pea-blossom becomes a pod containing peas; that the centerpiece of a peach-blossom becomes a peach, etc. But
probably you have never thought of the other and more
conspicuous parts of the flower as helpers in the work of
making seeds. Nor have you thought of the flower parts
as leaves changed in form and color to fit them for this
work so different from that of an ordinary leaf.

Take one of the flowers which you have brought to
study. Note the number and shape of the outside leaves.
They are *sepals*. Next to the sepals are larger and more
showy leaves called *petals*. The slender stems with heads
are *stamens* and the centerpiece is called the *pistil*. The
flower then is made up of four kinds of organs.† If
you have a flower with the sepals all alike, and the petals all
alike, and not united, remove those organs. You have now
a fair view of the stamens. The head of a stamen is called

---

* Wind-fertilized flowers are prodigal of pollen, while those most perfectly
fertilized by insects waste little or none. Burs get free transportation everywhere; berry seeds are carried by animals to their chosen fields.

† The leaves of each set may be united for a part or all of their length. In
eschscholtzia the two calyx leaves are completely united and the pair forming the
pistil are joined for part of their length. Two or more leaves of a set may be
united and the others distinct. In pea-blossoms nine stamens have the filaments
united for two-thirds of their length, and one is distinct, two of the petals are
united, and all the sepals.

an *anther* (see the figure of a mustard flower), and the stem a *filament.* Find on the anther *pollen* grains. The pollen grows inside the anther. Note that the anther consists of two sacs or cells which open to let the pollen out. In a bud the anther is full and in an old flower it is empty. Remove

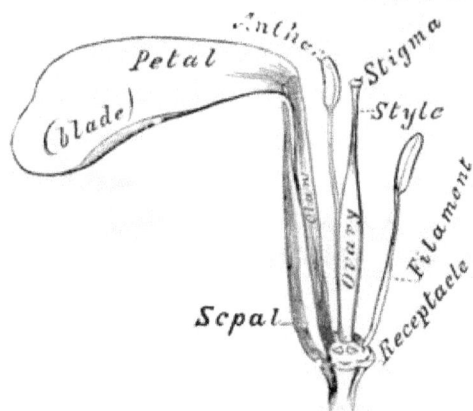

the stamens. The pistil has an enlarged base called the *ovary*, a neck or stem called the *style*, and one or more *stigmas* at the top. In a mustard flower there is one stigma; in a geranium five; in an eschscholtzia there are really but two, each divided into

Fig. 49. Mustard flower magnified with four of the stamens, three petals and three sepals removed.

several slender parts; and in any wild morning-glory there are two stigmas. Cut across the ovary and find the *ovules* (the beginnings of seeds). If you have eschscholtzias or morning-glories Fig. 50 will help you in studying them. Bring for the next work a bunch of geraniums.

Exercise 36.

Study of a Geranium.—The sepals of a flower taken together are called the *calyx.* Note that the upper sepal is larger than the others. The petals form the *corolla.* Like the calyx the corolla is slightly irregular or one-sided.

Two equal petals on the upper side are smaller than the others and often slightly different from them in color. Find between these petals a small hole extending down into the stem. Split the stem to find its depth. Taste the liquid in it. It is *nectar* and the hole is a *nectary*. Do you think a bee could get the nectar? Would the difference between the upper petals and
the others make
it easier for an
insect to find
the nectar tube?
Evidently gera-
aniums prepare
nectar for cer-
tain insects and
make it easy for
them to find it.
Were the corol-
las green would
it be so easy to
see them at a

Fig. 50. *a.* Bud of eschscholtzia with the removed calyx-cap above. *b.* Flower of the same with two petals removed, one of which is seen below bearing stamens at its base. *c.* Field morning-glory, the petals completely united. Above the corolla is split to show the stamens.

distance? Is their color helpful to the insects? Is it not reasonable to suppose that the visits of these insects are beneficial to the flower? In order to understand how insects can help flowers make seeds you must know more about the work of stamens and pistils. Many observations and much experimenting have proved that ovules cannot become

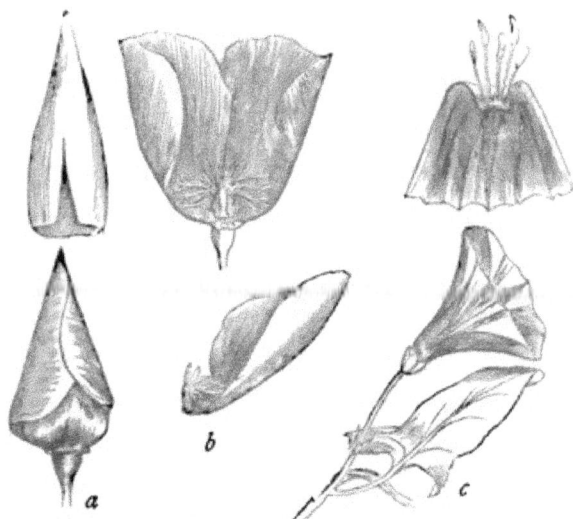

seeds unless pollen from the same flower or another flower of the same kind reaches the stigma. The flower is then said to be *fertilized.** It has also been shown that usually the best seeds are produced when the flower has been fertilized by pollen of another flower. This is called *cross-fertilization.* When stigmas receive the pollen of stamens which are with them in the same flower we say the flower is *self-fertilized.* Cross-fertilization is chiefly brought about by insects; but grasses (wheat, corn, barley, etc., are in the grass family), date-palms and most of our forest trees are fertilized by pollen carried in the wind. The flowers of such plants, having no use for signals, are without petals, *i. e., apetalous.* Indeed, the flowers of oaks, poplars, and pines are so inconspicuous that many people think these trees are flowerless. A few inconspicuous flowers need the help of insects. These bear nectar and attract insects by their odor. Highly colored flowers and fragrant flowers usually bear nectar, and are so constructed that their insect visitors in getting the nectar must rub against anthers when they are shedding pollen, and against stigmas when they are ready to receive pollen. Thus, in the common hollyhock, when the anthers are covered with pollen, the stigmas are hidden in a tube formed by the filaments, and after the pollen is shed the stigmas come out. Insects vis-

---

* Pollen grains on the moist stigma send sprouts down through the style into the ovary, where they come in contact with the ovules. Ovules not thus reached do not grow. If the school is provided with Barnes' dissecting microscope the shape and even the markings of pollen grains may be shown by combining several lenses so as to reduce the focal distance to less than half an inch.

iting flowers in the first condition become dusted with pollen which they rub off on stigmas of flowers in the second condition.

Examine the flowers in your bunch *(umbel)* of geraniums. In what condition are the stigmas of the flowers just opened?* Examine a pistil in a flower which has shed its petals and stamens. Note five projections at the base. Examine still older pistils if you have them. A geranium pistil is evidently made up of five parts. These parts are carpellary leaves or *carpels*. Bring sweet-peas to-morrow.

EXERCISE 37.

Study of Sweet-peas.—Hold the peduncle in a natural position. Do the flowers face away from the plant on which they grow? How should they face to make them most conspicuous? Show how such a position would also make them more convenient for the insects who are invited to sip their nectar. Where do you think an insect would alight? Observe that the upper petal is erect (or, as in some forms, projects like a canopy over the others), and that the side petals are nearly horizontal, covering the lower pair, which are united and curve upwards. Which of the side petals overlaps the other? Hold a flower firmly by the calyx and press downward upon the right-hand

---

* Without a good lens it is not easy to tell when the anthers are shedding pollen, but it is evident that at first the closed stigmas would not receive pollen, for the outer surface is not stigmatic (naked and sticky). Double geraniums are monstrosities and teach us nothing concerning fertilization. If possible, study wild flowers, particularly those with irregular corollas.

petal with your pencil. Try the left petal, holding the pencil horizontally in the channel on the upper side. Does the style then push upwards through the apex of the pouch formed by the lower petals? Which side of the style is hairy? What do the hairs brush out of the pouch? Try to estimate the weight of an insect which resting on the petal would cause the style to protrude. Is a honey bee heavy enough? Where would the pollen brush rub against the insect? Instead of pressing with the pencil upon the left petal only, let the point pass obliquely across to the right petal. Do you think an insect would be likely to place some of its feet over on the right side? Would that make the style with its pollen-brush work better? Is the weight of the insect the only power which it can use to bring the petals down? Would an effort to push up the base of the erect petal add to the downward push of his weight? Would such an effort help an insect to get at the nectary? Where is the nectary?

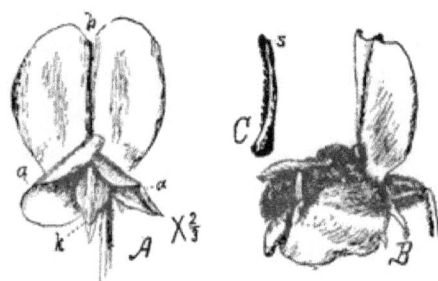

Fig. 51. *A.* Face of a sweet-pea. *a a.* Side or wing petals, the one on the left higher and overlapping the other at the apex. *b.* The upper or *banner* petal. *k.* The lower pair of petals united, forming the *keel*, which encloses the stamens and pistil. *B.* Shows how a bumblebee mounting the wings on the left side brings them and the keel down, thus causing the style to brush pollen upon his right side as with the petals he sinks to the level of the unyielding ovary. *C.* A magnified view of the style as seen from the front, showing the brush of upturned hairs which sweeps the pollen out of the pocket formed by the tip of the keel. *s.* The small stigma.

Look out for insects among the sweet-peas to-morrow morning. Save your sweet-peas or bring more for the next exercise. Each pupil should have a peduncle bearing two flowers and a bud; or, one flower and two buds.

EXERCISE 38.

Study of Sweet-peas Continued.—Have you observed insects visiting sweet-peas? Were any of them collecting pollen with their hindermost legs? Did those seeking nectar receive any pollen? Observe again an open flower. Do you know whether the petals have the same position at night as in the daytime? If the upper petal remains erect in a rain storm could water enter the nectary? Of what use are the two projections at the base of the erect portion of the upper petal? Remove the upper petal and the side pair. Cause the style to go in and out of the pouch that you may see how it works. Which way do the hairs of the pollen-brush point? Explain how it works. Only the tip of the style is stigmatic. Take one of the buds. Note that the upper petal encloses all the others. Remove it and the side pair. Is there any loose pollen in the pouch? Hold it up to the light. Observe that the lower and front edges of the two petals which form the pouch are united from near the base to the tip of the enclosed style. Loosen the pouch at the base and slip it off without injuring the stamens. Count the stamens. With a needle or pin raise the distinct one which is above. The others are united for the whole length of the ovary or two-thirds of their length. Find the nectar at the base of the sheath formed by these united filaments. Do you see how an insect can get the nectar? Observe the stigma. Is there any adhering pollen? If the stigma is not ready to receive pollen before the pollen of its own flower has become useless, it is evident that fertilization cannot take place without the aid of insects heavy enough or strong enough to cause the protrusion of the style. In any event there would be some cross-fertilization.

Make drawings to show this mechanism of the flower.

Exercise 39.

[Read and discuss in the class.]

Flowers and Fruits.—Keep constantly in mind when studying a flower the following statements and be on the alert to discover evidence to prove or disprove them: The work of flowers is to produce seeds. Insects usually, though unwittingly, assist in this work by carrying pollen from the stamens of one flower to the stigmas of another of the same kind. That they may be induced to do this work the flowers secrete nectar and serve it in such dishes (nectaries) that only the invited guests can get it. When the feast is spread the fact is signaled to the insects by the unfolding of the corolla. The parts of the flower are so shaped and arranged that each guest must rub against anthers when they are discharging pollen, and against stigmas when they are in the proper condition to receive pollen. The receptacle rarely does more than hold in place the real organs of the flower. Sometimes, as in strawberries and pineapples, it forms a part of the fruit. The calyx serves to protect the other organs while they are tender in the bud and usually continues to protect the ovary while it is developing into fruit. Often it forms the outside of the fruit, as in apples, pears, pomegranates and squashes. The corolla is a signal which shows insects where the feast of nectar is spread. It also keeps unwelcome visitors out and helps the welcome ones to find and easily get the nectar, while it assists stamens and stigmas to deliver and receive pollen. The work of the stamens is finished when the pollen which they have produced has been delivered to the insect messenger. The pistil (carpel or carpels), usually dismissing the stigma soon after the pollen has been received, continues its work until it becomes fruit containing ripe seeds. Fruits are green in color until ripe, then they usually change to a color that makes them more easily seen. The pod or juicy outside of fruit usually causes the seeds to be scattered often many miles from the plant which produced them. Cockle-burs, bur-clover, etc., cling to animals and are thus distributed. The sweet pulp of berries causes them to be scattered by birds, etc. In short, an apple is good that its seeds may be distributed. A cherry is red that some cherry loving animal may surely find it and drop its seed far from the parent tree.

EXERCISE 40.

Tertiary Organs or Changed Secondary Organs of Plants. *Rudimentary and Obsolete Organs.*—Your attention has already been called (page 38) to the fact that when an organ ceases to be useful it becomes *rudimentary*, and finally entirely disappearing it is said to be *obsolete*. For example, when flowers grow in clusters, to form larger signals, and to make it easier for their insect aids, thus summoned, to gather nectar, the subtending leaves doing more harm than good are reduced to bracts, or become obsolete, as in most plants of the mustard family. When an improved condition of a flower lessens the amount of pollen required, one or more stamens become rudimentary or obsolete. Thus, in sage, as shown in Fig. 52, by an ingenious modification of two stamens they are made as effective as were the original five. The useless ones are obsolete.*

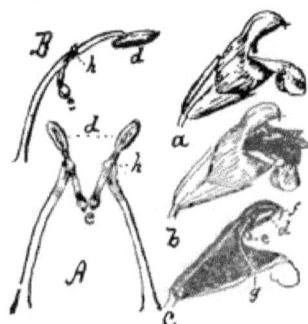

Fig. 52. Mechanism for securing cross-fertilization in the common garden sage. *a.* The flower as it appears when the anthers are ready to give up their pollen. The stigma not yet ready to receive pollen shows its two lobes protruding from the tip of the upper lip of the corolla. *b.* A flower visited by a bee, who receives the pollen on his back while sipping nectar from the bottom of the corolla. *C.* Shows the position of one of the pair of stamens which stand at the mouth of the corolla. *d.* The anther cell which yields pollen. *e.* The abortive cell, which is pushed backward by the bee as he enters, thus bringing the other cell down upon his back as shown in *b.* *f.* The stigmas. *g.* The base of the filament. *A.* Shows the stamens as seen from the front. *B.* A side view of one stamen. It is plain that pushing *e* to the left would cause *d* to descend by turning on its joint at *h.*

Promoted and Degraded Organs.—In a few cases when inconspicuous flowers are gathered into heads some of their subtending leaves, instead of being degraded to the condition of bracts, are promoted to fill the office of petals. Thus the corolla-like involucres of dogwood and anemopsis are formed (Fig. 80).

* In some native sages two of the three upper stamens are rudimentary; but the middle one directly in the way of the style is obsolete. In most irregular flowers only the upper stamen is rudimentary or obsolete.

Often, particularly in dry countries, branches or leaves become protective spines. In our hookeras (Fig. 64) the three sterile stamens (staminodia) put their heads together for the purpose, probably, of keeping out intruders, as well as to insure the economical collection and transference of pollen. In many cultivated plants stamens no longer necessary for the production of pollen cease to attempt usefulness in that way and develop into beautiful petals to please man.*

Questions That Should be Considered in the Discussion.— What parts of a flower are modifications of the stem? Which of the four kinds of leaf parts are most modified or changed from the ordinary leaf form and nature? Which are the least changed? How does a nasturtium leaf do the work of a tendril? What materials are stored in a beet? In a potato? What harm would be done by ordinary leaves in place of the bracts in a bunch of lilac blossoms? Is a spine doing a lower work than a leaf? Explain how man can change the organs of a flower or the character of fruit?

Exercise 41.

[Read and discuss in the class.]

The Growth of Fruits, Seeds and Embryos.—The development of fruit from the ovary is often very interest-

---

* Man is, perhaps, the only animal excepting insects, and a few birds, that has had any influence upon the arrangement, size, form, color, and odor of flowers. Doubtless the present condition of many wild fruits was brought about by the higher animals. Man, beginning where these animals left off, has greatly modified the fruits of plants which he cultivates.

Some plants would long ago have become extinct had it not been for man. Indian corn and wheat, for example, are not known in a wild state.

ing. You should daily observe the growing pistils of several plants from the time the corolla falls till the seeds are ripe. At intervals of a few days one of the young fruits should be picked and dissected. If you follow the growth of an almond you will find it has at first two ovules (one for each edge of the carpellary leaf). Usually one soon dies and the almond ripens but one seed. In the very young acorn there are the beginnings of six seeds (two for each carpel) but only one grows. So also in buckeye usually but one of the six ovules grows. In an early stage of its growth the young buckeye fruit appears as shown in the figure at *a*.

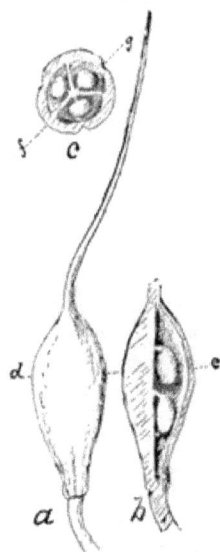

Fig. 53. *a*. Growing pistil of a buckeye. *b*. Vertical section of the ovary cut in the plane *f g*, shown at *c*, which is a cross-section cutting the ovary in the plane *d e*, as seen in *a* and *b*.

Fig. 54. A diagram showing the one-sided development of the buckeye after it has reached the size of a walnut. The midribs (dorsal sutures) of the carpellary leaves are marked *m*, and their lines of union (ventral sutures) are marked *s*. *a*, *b* and *c* are the ovules, two in each cell. The fertile carpel makes about half the pod—the lower half in the figure, from *s* on the left to *s* on the right.

The cross-section *c* shows the three cells and the vertical section *b* shows how the two ovules of each cell grow one above the other. Fig. 54 is a diagramatic section of the condition when the fruit is nearly grown. Five of the ovules grow to be as large as mustard seeds or larger and then stop growing. The sixth one

becomes a large seed and crowds the partitions of the pod to
one side along with the abortive ovules. Notice that the
carpels to which the undeveloped ovules belong do not
grow as much as the other one, thus making the fruit one
sided as shown in the drawing. The carpels split along
the midribs (*m, m, m*) when the pod opens. The midrib
of the fertile carpel splits smoothly, but the other two are
partly held together by the partitions which have been
forced against them. Do not fail to examine these fruits
and their seeds. Probably buckeyes once—thousands of
years ago when the climate was different—ripened all of
the ovules into small seeds.

Exercise 42.

The Growth of Ovules.—You cannot study the
development of embryos from the beginning, without the
help of a compound microscope, but you can easily observe
all stages of growth from a tiny green speck to the full-
grown embryo. Most seeds are nearly full grown in
appearance before the embryo is more than fairly visible to
the naked eye. The seed coat, filled with a syrupy or
milky, usually sweet, liquid, appears to constitute the very
young seed. With a sharp knife cut in halves a great many
green-peas, in size from half-grown upward. You will
surely find in some of them tiny green embryos, and you
may get specimens from the size of a pin's head up to those
which tightly fill the seed coat. At the top of the figure is
seen—magnified two diameters—the young seed of a lupine,

cut so as to show the growing embryo lying in one end. In the same figure is represented a radish pod, laid open so as to show three of the seeds, two of which exhibit their partly grown embryos.* Below, at *b*, is one of these magnified, and at *a* an older one, also magnified. The grown embryo completely fills the seed. Observe the positions of the embryos in relation to the stems of the seeds and the stems of the pods. The lower seed in the radish is fastened to the lower side of the pod, the middle seed grows to the upper side. The cotyledons increase much more in size than the caulicle. The embryo evidently grows, in part at least, by absorbing the liquid around it. Suppose the embryo of the lupine to quit growing at the size represented in the figure, and that the liquid around it thickens until it becomes solid. Would not the seed thus formed be endospermic?

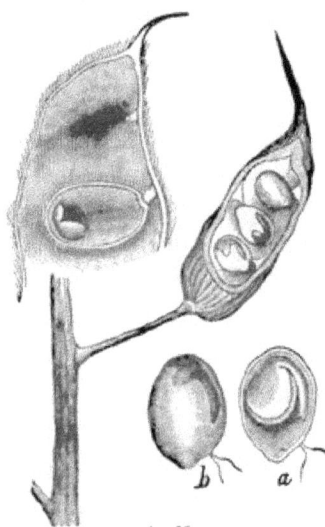

Fig. 55.

Take the pods which you have brought and study them as indicated in the preceding paragraph.

Get many kinds of flowers for the next exercise.

---

* These are cut in two. The embryo may be seen through the seed-coat, as represented at *b*, by holding it up to the light. Half of the seed-coat is removed from *a*.

EXERCISE 43.

Cohesion of Floral Organs.—When organs of the same kind grow fast to each other they are said to *cohere*. *Adhesion* is the union of unlike parts. For example: the petals of a petunia cohere, and the stamens adhere to the corolla. The extent of the cohesion varies from a slight union, usually at the base, to a complete union from base to apex. In most flowers the carpels are united to form a compound pistil; in many the sepals cohere; in nearly as many the petals; and in a much smaller number the stamens are united. The united portion of calyx, corolla or stamens is called the *tube*. The upper end of a corolla tube is the *throat*. The distinct upper parts of sepals or petals form the *lobes*, which in the calyx are often called *teeth*. The border of a calyx or corolla is called the *limb*, which is *entire* if the union of the floral leaves is complete; as in morning-glories and petunias.

In describing a flower, it is not necessary to tell directly whether the sepals or petals are united or distinct. When we say that a pink has five long-clawed petals the statement implies that they are separate. Were they united the description would be: Corolla limb five-lobed, etc.*

---

* Pupils in botany are often required to memorize many useless terms. Such words as *polypetalous* or *choripetalous*, *gamopetalous* or *sympetalous* are not worth their memory cost. They add neither to the conciseness nor the accuracy of a description. "Petals united" is shorter than "corolla gamopetalous." These words and many others are relics of the time when all botanists wrote their descriptions in Latin. The words *choripetalæ*, *sympetalæ*, etc., are convenient headings in systematic botany.

Fig. 56. Œnothera ovata. *s*. Surface of the ground. *r*. Rootstock from which grow the leaves and sessile flowers. No stem, not even a peduncle, appears above ground, and the pods ripen underground. The style and its coat, the cohering calyx tube makes the apparent flower stems.

The word corolla, then, need not be used in the description of a flower which has distinct petals; and the word petal is not necessary when the petals are united.

Classify your flowers by putting in separate groups (1) those with distinct petals and (2) those with united petals. Note the cohesion of the sepals. Take in turn each of the flowers in which any of the organs are coherent and try to discover whether the union is of any use in the work of making seed. Find the nectaries. Have any of your flowers slender tubes? Decide in each case what kind of an insect would best assist in cross-fertilization. Are the carpels distinct in any of your flowers? Are they completely united in any? Describe one of the flowers. Your description of a mallows blossom ought to be like this: "Mallows flowers are half an inch or less in breadth on peduncles less than an inch long. The five-lobed calyx bears three bracts near the middle. The pinkish petals are held together at the base by the numerous stamens which growing upon the claws of the petals cohere to form a tube. This tube is surmounted by the distinct upper ends of the filaments bearing kidney-shaped anthers. The eight or more carpels are crowded tightly together and slightly cohere around a central projection of the receptacle, their thread-like styles filling the tube formed by the stamens."

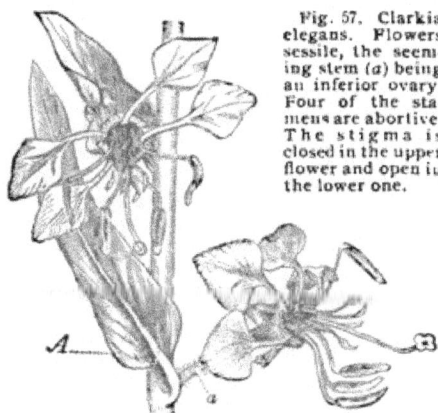

Fig. 57. Clarkia elegans. Flowers sessile, the seeming stem (a) being an inferior ovary. Four of the stamens are abortive. The stigma is closed in the upper flower and open in the lower one.

Get for the next exercise as many kinds of regular flowers as possible with the parts in threes. Bring also if you can fuchsia, and other flowers with the parts in fours. Perhaps you can find an œnothera* like that shown in Fig. 56; or a clarkia (Fig. 57).

EXERCISE 44.

Adhesion or Adnation of the Floral Organs.—In the flowers of endogens, which usually have their parts in threes, the calyx and corolla are often made up of leaves

---

*This cowslip œnothera (pronounced, e-no-the'-ra the *th* soft) is common near the coast. A smaller species is found in the interior. Children in picking these flowers usually leave the ovary underground. If the class is studying in September or October, rose-colored godetias (gode'tia) or Zauschneria (Fig. 59) may be found. Squash or melon flowers both staminate and pistillate are desirable.

nearly alike in size and shape, and exactly alike in color. Frequently the sepals and petals are united by their edges to form a tube or a cup. Since, in such cases, the two sets of organs are together doing the same work—that usually done by corolla alone—it is convenient to give them one name. We call such flower cups *perianths.* The adherent or adnate portion is called the *perianth tube;* and the free parts, the *perianth lobes.* When the perianth divisions are united, the stamens are adnate to the perianth

Fig. 58. Perianth of triteleia laid open, showing the pistil on a stem and the adherent stamens.

tube, as shown in the figure. In flowers with coherent petals the stamens are adnate to the corolla. In many flowers the calyx is adnate to the pistil, forming the outer coat of that organ for a part of its length. The extent of this adnation varies. In some flowers the calyx is adnate to only the base of the ovary; in others the entire ovary is thus covered, and in a few it is united even to a long style for nearly its whole length, as is shown in cowslip œnothera (Fig. 56). Since the ovary seems to be under the flower when the calyx is adherent, it is usually described as *Ovary inferior.* It is not always easy to tell whether the ovary is inferior or superior.

Sometimes what seems to be a peduncle is an inferior ovary (Fig. 59); or, as in the œnothera, an inferior style also. A vertical section, as shown in Fig. 60, will generally decide the question. When the ovary is inferior the corolla and stamens must be adnate to the calyx or—and this rarely happens, except in orchids—to the ovary or style.

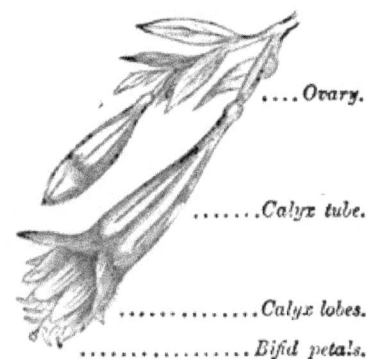

.... Ovary.

.... Calyx tube.

.... Calyx lobes.

.... Bifid petals.

Fig. 59. Ovary inferior and stemlike. Calyx petaloid.

Pick out all the flowers which manifestly have inferior ovaries, then find the ovaries in the other flowers, so as to be sure none of them are masquerading as peduncles. Also make vertical sections to detect any cases of ovaries partly inferior. Make drawings showing vertical and cross-sections of the ovaries. The latter will usually show the number of carpels. The ovules grow in as many distinct sets as there are carpels. Sometimes the sets are in separate cavities called *cells* (Fig. 52); and sometimes they are on the sides of one central cell (Fig. 61*a*); or, rarely, they grow in the center of a one-celled ovary (Fig. 61*b*). Remembering that carpels are leaves, which develop so as to form closed spaces or cells, it

Fig. 60. Ovary partly inferior.

becomes easier to understand the structure of pistils. If there is but one carpel, as in the flowers of peas and peaches, the midrib appears as a ridge or channel along the lower side of the pistil and the inturned edges make a different mark—usually depressed—along the upper side. Inside, upon the edges of the leaf, grow two or more ovules.* A compound pistil may be formed by several such folded carpels joined with the midribs outside. Such a pistil will have as many cells as carpels (Fig. 52), and there will usually appear twice as many longitudinal marks, representing, alternately, the midribs and the lines of union. Sometimes, as in violets, the carpels are not folded, but joined by their edges to form a single cell, with ovules along the seams on the walls of the ovary (Fig. 61a). More rarely the ovules appear on a growth from the center of the base (Fig. 61b). Probably this is an outgrowth from the leaf edges at their bases.

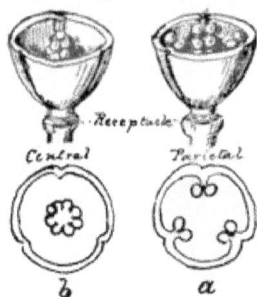

Fig. 61. Diagrams showing two 2-carpelled pistils; one (a) with the ovules on the sides (placentæ parietal); the other (b) with ovules on a stem in the center (pacentæ central).

Save any flowers you have left and bring more for the next exercise. Secure also many kinds of flower buds.

EXERCISE 45.

Arrangement of Floral Leaves in the Bud.—Sepals and petals in the bud are usually arranged in one of the

---

* We would expect the number to be even—as many ovules on one edge as on the other. Though apparently there is but one row of ovules in the carpel of a pea, there are really two, as may be seen by splitting the pod. Stone fruits usually have but one seed; but there are two ovules. Sometimes there are two or three carpels in a peach blossom, and often the Hungarian prune is double.

three ways shown in Fig. 62. When the edges meet without overlapping they are said to be *valvate;* when one edge is in and the other out all the way around, giving a twisted appearance, they are *convolute;* and when they overlie each other like the leaves of a cabbage head, they are *imbricate.* Petals are sometimes crumpled in the bud as in the true poppies. In morning-glories and some other flowers which have the petals united so as to form an entire or nearly entire limb the corolla is plaited in five folds which overlap each other convolutely. In pea blossoms, as we have observed, the lower pair of petals cohere along one edge forming a sheath which encloses the stamens and pistil. These in turn are covered by the side petals and the whole is enfolded by the upper petal. In many irregular flowers the three lower lobes fold up over the upper pair. Filaments and styles are sometimes doubled or coiled in the bud, as in eucalyptus and four-o'clock.

Fig. 62. Floral leaves in the bud. *a.* Young bud of lavatera with valvate sepals enclosing the rest of the flower. *b.* An older bud of the same, the convolute petals ready to open.

Make cross-sections of your buds and draw diagrams to show the arrangement of the parts. As a map shows better than a picture the relative positions of features in a landscape, so a diagram of a bud or flower shows more clearly its plan than a picture or even an accurate drawing of a section. A careful study of Fig. 63 will show you this fact and enable you to make similar diagrams. Do this.

Provide a fresh supply of flowers and buds for the next work. Periwinkle, oleander (single), pansy, potato, nightshade, manzanita, madrona, are desirable.

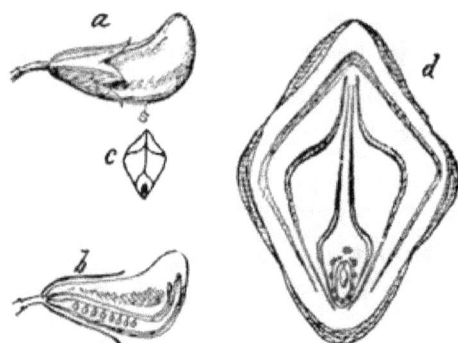

Fig. 63. Bud of sweet-pea shown natural size at *a*.
b. A longitudinal section. *c*. A cross-section at *s*. *d*. A diagramatic section representing the relations of all the parts to each other in the bud. An ovule is in the center surrounded by the carpel. The next enclosing line represents the stamens all united except the upper one. Then follow the two keel petals united below; the wing petals not united and the upper petal (banner) enclosing all. Outside are the five united sepals; one below, two above, and one on each side.

Exercise 46.

Stamens.—In a typical flower—the ancestral form, probably, of all complete flowers —the stamens, like the other floral leaves, grow separately upon the receptacle. As you have learned, they often seem to grow upon the calyx or corolla, and sometimes upon the pistil. Since it is probable that in all cases they originate in the receptacle, it is better to say they are adherent or adnate to the calyx, corolla, or pistil.* When, as in hookera (Fig. 64), the filaments are easily traced down to the receptacle, the stamens may be described as inserted at the summit of the perianth tube, with filaments decurrent (running down) to the base, the free portion erect and shorter than the anther; or,

---

* Perhaps an inferior ovary is covered by traces of stamens and petals beneath the calyx coat.

describing in another way those of a similar flower: the stamens of triteleia (Fig. 58) are in two sets; those opposite the outer segments (sepals) adherent for two-thirds the length of the perianth tube, and the others adherent up to the summit; the free portions about equalling the versatile, introrse anthers.

Stamens vary even more than carpels from the typical leaf. The hypothesis that the filament corresponds to the petiole, and that the anther is a modified blade, the lobes homologous with the right and left sides, seems reasonable, but good authorities consider the anther to have more the nature of an appendage or gland-like outgrowth of the leaf.

Remove one stamen from each kind of flower, noting which side is next to the pistil. Place them in a row, with the inner sides on the left. If there are any with anthers not two-lobed, put them at one end of the line. Next separate those which have the filaments attached to some point between the ends of the anther, as shown in Fig. 58.

Such anthers are *versatile.* The others are probably all

Fig. 64. Flower of hookera and perianth split to show stamens and staminodia; the latter without filaments continued to the base. *d.* A bud. *f.* The pistil.

adnate: that is, the anther grows fast to either the inner or outer side of the filament, which seems to form the back, or nearly flat side, of the anther. If the filament seems to lie between the anther lobes (cells), so that the sides are nearly

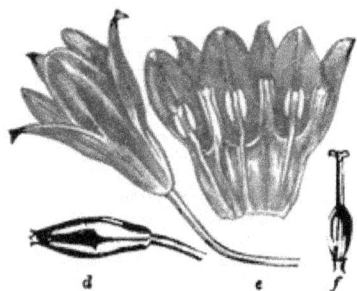

or quite alike, the anther is *innate*. Observe carefully how the anther cells open to let the pollen out. Which open by splitting lengthwise along the face? Which by splitting along the edges? Do any discharge the pollen through holes at the top? Some stamens face the pistil (inward). These have *introrse* anthers. Those with their backs to the pistil are *extrorse*. Take a stamen from each kind of bud and compare with one from the flower. Have you any flowers with two kinds of sta-mens? Are any sterile? Draw all those taken from the buds, representing a profile and a face view of each kind. Also draw a few which are discharging pol-len. Do any discharge pollen in the bud? Label your draw-ings in this way: (*a*) Mustard stamen—one of the long ones —in the bud. Anther adnate and introrse. (*b*) Same in open flower. Anther extrorse by a quarter twist of the filament; pollen escaping by slits. Why do the four long stamens of mustard—standing as high as the stigma—turn so as to face away from the center? The short pair have introrse anthers in the bud, and they remain facing the pistil in the open flower. Why do they not also turn away from the pistil? Do you find any anthers with four cells? Are there any stamens with appendages to the filaments or the

Fig. 65. *A. a.* **Flower of a common** shrubby lupine. *b.* The same with the upper and side petals removed showing the united pair of long-clawed petals which enclose the united stamens and pistil. *B. a.* **Same with all the petals** removed **showing 5 empty anthers.** *b.* Stamens as they appear in the bud, with the anthers all full. The short stamens of the bud become the long stamens of the flower. *c* and *d*. **Anthers** magnified.

anthers? Sometimes the filament continues on above the anther, as in violets. Have you any reason to give for the peculiar behavior of stamens shown in Fig. 65?

Get flowers of as many different shapes as possible for the next work.

EXERCISE 47.

The Forms of Flowers.—The shape of a flower depends chiefly upon the forms of its parts, and their adhesion or cohesion. Since the corolla is usually the largest of the four sets of organs, its form mostly determines that of the flower; but sometimes, as in fuchsia and zauschneria, the calyx has more to do in shaping the flower than the corolla* (See Fig. 59). It is, as it were, the chief color-bearer, for it is as highly colored, and its parts, being united, are more highly developed than those of the corolla. In lilies the sepals play an equal part with petals in signaling insects, and in shaping the display of color. Clematis having no corolla, shows white or purple stars made of sepals only. When a calyx thus helps or does alone the work of a corolla it is said to be *petaloid*.

Unequal petals, or petaloid sepals, form an *irregular* flower. Such flowers are usually one-sided and bilaterally symmetrical; but when the parts are of even numbers, as in dicentra (commonly know as bleeding-heart), the flower may not be one-sided. The shape as well as the position of a

---

* Rarely stamens or pistils are the conspicuous organs of a flower. Then they are often petaloid as to color at least. The calistemons or bottle-brushes of Australia are such flowers. In canna, which should be examined if possible, sepals, petals, stamens and pistil are all petaloid and nearly exactly alike.

flower often depends greatly upon light, heat, and moisture; or, in other words, upon the time of the day and the kind of weather.* In describing flowers we should give the positions and forms when most actively at work, with banners widely spread, nectaries open, and stamens ready to give pollen, or the pistils waiting to receive it.

The form of a flower depends upon the inclination

Fig. 66. Silver bells (Calochortus albus).

of its conspicuous parts— or the lobes of these when united—to the axis or peduncle, as well as upon their shape and adhesion or cohesion. For example, the sepals of clematis and the corolla lobes of trientalis, spreading at right angles to the floral axis, form wheel-shaped or *rotate* flowers; the perianth of crown-imperial

Fig. 67. Regular corollas with united petals. *a.* Tubular corolla of tree-tobacco (*Nicotiana glauca*). *b.* Salverform corolla of slender tobacco (*Nicotiana attenuata*). *c.* Rotate corolla of solanum. *d.* Campanulate corolla of blue-bells.

or of our most common fritillaria and the corolla of cam-

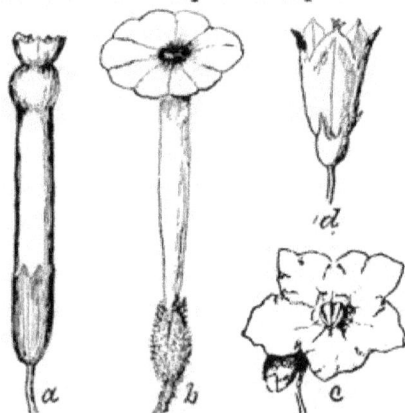

---

* Some flowers are open at night only. Most flowers droop and partly close in the rain. Pupils should study the behavior of plants in their homes out of doors. Often the appearance of a plant changes greatly after it is plucked and brought into the house.

Some of the resting positions or shapes of the floral organs are interesting. The corolla of a morning-glory, for example, when it quits work covers the stigmas with a cap very unlike that which crowned the bud. And there is a reason for this.

panula, with leaves spreading but curving toward the axis, make *campanulate* flowers; the recurved perianth of a tiger-lily looks like a turban; the incurved one of a common white calochortus is nearly globular or ovoid (Fig. 66), and the strongly infolded petals with the widely spreading sepals of another yellow species give a remarkable form.

a. *Upper lip or galea.* b. *lower lip.*
—*Aphyllon.*
Orthocarpus.

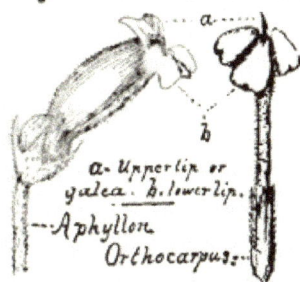

Fig. 68. Bilabiate corollas.

Fig. 67 shows common forms of corollas with united petals. A corolla or calyx is *tubular* if its parts are united nearly to the ends into a slender scarcely widening tube with erect lobes. If the limb is rotate and the tube slender so as to make the flower look like a salver it is *salverform.*

Irregular flowers usually have the upper pair of petals different in shape and size from the side pair and the lower one. When the petals are united the upper pair forms the *upper lip* and the other three the *lower lip* of a *bilabiate* corolla.

Fig. 69. Bilabiate corollas having elongated galea. *a* and *b* have the l wer lip reduced to three tooth-like projections. *d.* The corolla and *c* the entire flower of a rose-purple orthocarpus.

Figures 68 and 69 show some of the many hundreds of forms of bilabiate flowers.*

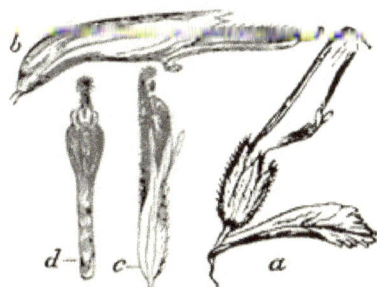

---

* Orchids, of which there are nearly three thousand kinds, mostly natives of the tropics, all have irregular flowers. One petal called the lip is often very different from the others in size and form (Fig. 70). All are adapted to insects, and only a few can produce seed without their aid. There are on this coast less than thirty kinds, mostly with small flowers.

Classify your flowers and describe their forms. Try to decide what must be some of the characteristics of their insect aids. Get as many kinds of flower clusters as possible for the next exercise.

EXERCISE 48.

The Forms of Flower Clusters.—You have already learned that flowers are either at the ends of stems or between leaves and stems; that is, they are either terminal or axillary. Flower clusters are also terminal or axillary, and the flowers in the clusters are terminal or axillary in relation to the peduncle or axis of the cluster. Often a terminal cluster is made up of flowers which grow in the axils of bracts; that is, the inflorescence is axillary. Sometimes an axillary cluster is composed of flowers which are terminal.

Fig. 71 represents forms of simple axillary inflorescence. Shortening an ordinary stem with axillary flowers and reducing the leaves to bracts would give it the appearance shown at *a* in the figure. A *raceme*, then, is a stem with very short internodes bearing flowers in the axils of bracts. Note that the oldest flower

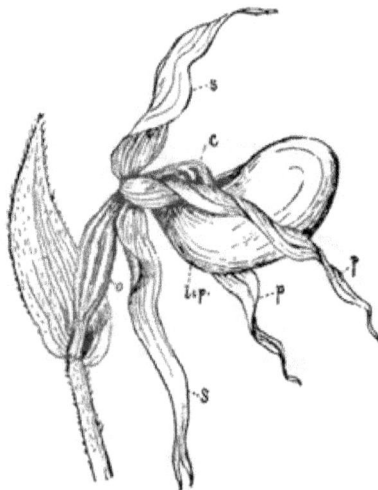

Fig. 70. Cypripedium, or moccasin flower. *s s.* Sepals; the lower pair united. *p p.* The lateral pair of petals. The lower one forms a sac which is the conspicuous part of the flower. *c.* The *column* formed by the united s amens and stigma. *o.* The inferior ovary.

is the lowest. Evidently the *spike*, shown at *b*, is a raceme of

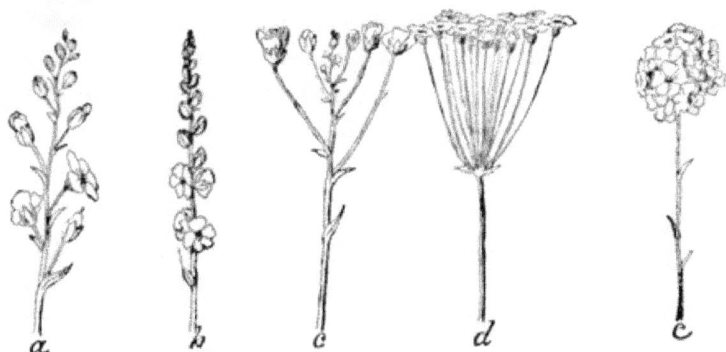

Fig. 71. Forms of Axillary Inflorescence. *a*. Raceme. *b*. Spike. *c*. Corymb. *d*. Umbel. *e*. Head.

stemless (sessile) flowers. When the peduncles of the lower flowers, by growing, keep pace with the central stem a flat-topped cluster called a *corymb* is formed. An *umbel* is formed when the internodes do not grow. The bracts then are all at the end of the peduncle and the pedicels belonging to them (subtended by them) all start from the same place. Shortening the pedicels in an umbel, thus bringing the flowers to nearly or quite a sessile condition, changes it to a *head*. The whorl of bracts in an umbel or a head is called an *involucre*. Racemes and spikes may be dense or close, loose or lax, drooping, erect, etc. A spike of apetalous flowers growing on a shrub or tree is called a *catkin* or *ament* (Fig. 72). The peculiar spike of a

Fig. 72. Catkins or aments of garrya. *a*. Pistillate flowers. *b*. Staminate flowers (lower end of a catkin three inches long).

calla, bearing in a consolidated mass yellow anthers above and greenish pistillate flowers below, is called a *spadix;* and the enclosing white leaf is a *spathe.* A raceme of racemes (compound raceme) is usually called a *panicle.* Corymbs and umbels are often compound.

Fig. 73 shows forms of terminal inflorescence. A *cyme* (*a*) resembles a corymb but the plan is very different as shown in the figures. Cymes frequently resemble umbels but are easily distinguished since an umbel has the buds or youngest flowers in the center. Sometimes terminal

Fig. 73. Forms of terminal inflorescence. *a.* Cyme of an opposite-leaved plant. *b.* A one-sided cyme of a plant with opposite leaves. *c.* Similar cyme of a plant with alternate leaves.

inflorescence takes the form shown at *b* and *c* in Fig. 73. At first view they seem to be racemes and are usually called *scorpioid racemes.* Usually in this kind of inflorescence the flowers are sessile and so densely crowded that they are thrown into two rows on the convex side of the stem which ends in a coil of buds as shown in Fig. 74.*

Fig. 74. Scorpioid inflorescence.

Study *a*, *b* and *c* in Fig. 73. Evidently *b* is the same as *a* with all the branches on the left removed

---

* On this coast, yellow or orange flowers in such *scorpioid spikes* belong to amsiuckias; blue ones, among wild flowers, are usually those of some kind of phacelia; and some krinitzkias have white ones, larger than those of the blue forget-me-not, which also grow in coiled spikes. Heliotropes have this inflorescence.

and the growth extended a few nodes farther. (Dotted lines show the positions of the lower two branches which are missing in *b*.) In *b* then only one of the pair of axillary buds at each node grew. The bracts are alternate in *c* and each axillary bud became a branch which produced but one ordinary internode and node (*pytomer*), the second internode being a peduncle ending in the receptacle which is a compound node. This reminds us of the grape vine, the upper portion of which is made up of branches each of which develops but one normal internode and then becomes a tendril (see Fig. 48, p. 61). Compare *b* in Fig. 73 with Fig. 75, which represents a pigmy poppy with the leaves in whorls of three. Only one of the three buds at each node becomes a branch, and at each node the main stem becomes a peduncle. Imagine all the stems—internodes—of *a* Fig. 73 to be shortened so much as to bring all the nodes together with their bracts in a whorl with the lower pair; and at the same time, the peduncles lengthened so as to bring all the flowers to a level at the top. Then

Fig. 75. Pigmy poppy or meconel-la (*Platystigma Californicum*).

the cluster would look like an umbel, but the oldest flower —gone to seed in the figure—would be in the middle, while in an umbel the youngest flower is in the center. When the pedicels are spreading such a cluster of flowers is called an umbellate cyme, but if long and nearly parallel

they form a *fascicle.* Very short pedicels reduce the umbellate cyme or fascicle to a *glomerule* which is like a head
in appearance.

As has been indicated, inflorescence may not only be compound but mixed. There are, for example, racemes of cymes
which resemble and are often
called panicles. Such a false
panicle is called a *thyrsus* or
thyrse. Buckeye and lilac blossoms are in thyrses. Racemes,
spikes, etc., are sometimes
bractless.

Classify your flower clusters
and describe them. It may be
necessary, as in describing
leaves, to combine terms. For
example: The cluster represented in Fig. 76 is between a
raceme and a corymb, and it is without bracts. We would,
therefore, describe it as a bractless, corymbose raceme.

Fig. 76.   A bractless corymbose raceme.

Bring for the next exercise any of the following or
similar flowers: sunflower, mayweed, chrysanthemum, aster,
marigold, dandelion, daisy, tarweed, thistle, lettuce, tidy-tips,
cosmos, marguerite, tansy, goldenrod.

EXERCISE 49.

Composite Flowers.—Take your largest specimen
and examine it. Do you find outside petal-like leaves simi-

lar to *a* in Fig. 77? Remove one of these *ray* flowers and
compare with the figure. Find the inferior ovary and the
pair of stigmas. Note that the ray is really a corolla with
petals united into a flat apparently simple petal instead of
into a cup or tube. Look for central or *disk* flowers similar
to *f* in the figure. Look for the peculiar calyx limb called
pappus. It may consist of scales or of hairs, and these
vary greatly in appear-
ance. Sometimes the pap-
pus is obsolete. Find in
the disk flower the stigmas
and the tube of anthers
which encloses them.
Note the appearance of
the involucre formed by
bracts outside the rays. Is
there a bract subtending
each disk flower? These
bracts called *chaff* may
be wanting, and usually
they are inconspicuous.

Fig. 77. Representing florets of tar-weed [See
page 4] and sunflower. *a.* A ray floret of tar-weed
and the enclosing bract which is covered with
tarry hairs. *b.* The bract enclosing the ripe akene.
c. Ray floret magnified, with all but the base of the
corolla removed so as to show the pair of stigmas.
d. Ray corolla. *e.* A disk floret magnified. The
disk florets are all sterile, producing only pollen.
f. Disk floret of sunflower. *g.* A ray floret of the
same. In sun-flowers the ray florets are sterile and
the disk florets are fertile.

Are any of your spe-
cimens made up entirely
of disk flowers? Of ray flowers? Classify according to
these differences.

Since these dense heads of flowers resemble single flowers they are called compound or *composite* flowers, and the
separate flowers are called *florets*. Draw some of the larger

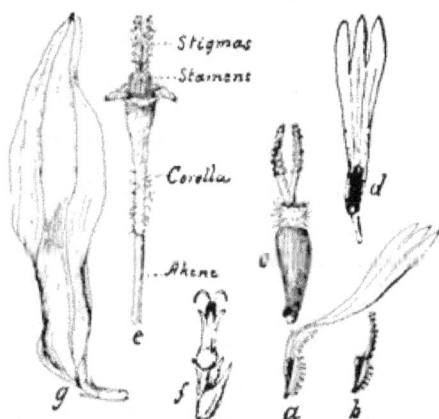

florets. Dissect out the stamens of a large floret. How many are there? Are the filaments united? Are the anthers extrorse or introrse? Are there any stamens in the ray flowers? What is the use of the rays? Of the disk corollas? Of the pappus?

EXERCISE 50.

[Read and discuss in the class]

Why Flowers are in Clusters and why a Composite Flower is the Highest Type of Inflorescence.— We have learned that a typical flower is a stem, or its upper part, in which only the lower internode, if any, is developed, and the greatly modified leaves of which are in four unlike sets.

In like manner we may consider a cluster of flowers to be a stem, with its branches bearing bracts and floral leaves. The lower internode of the stem becomes the peduncle, and that of a branch a pedicel. If all the internodes of the primary stem and the first ones of the branches are developed a raceme or a corymb is formed. This becomes a spike if the pedicels do not grow. If the internodes above the peduncle (*rachis*) fail to grow the raceme becomes an umbel which, in turn, if the pedicels are not developed, becomes a head. Evidently, then, a head, being the most modified, is the highest inflorescence of the axillary type, and in a similar way it can be shown that a glomerule is the highest form of terminal inflorescence.

Since plants or their parts are modified for their betterment, flowers in heads must be more effective than those in umbels, the latter than those in racemes, which, in turn, must surpass those widely separated.

Let us see if this is not reasonable. Suppose the flowers which form the umbellate cyme of our common garden geranium—more properly a pelargonium—were placed separately opposite the leaves. At once we see that the pedicels—now peduncles—would have to be considerably lengthened to bring the flowers distinctly into view above the leaves. This would be an additional expense to the plant. It is plain, also, that the scattered flowers could not be seen as far as the larger signal formed by the umbel. Fewer insects would visit them, and it would take them longer to go from one to another of the separated flowers than to pass from nectary to nectary of those in the cluster. In like manner it can be shown that flowers do better work in a head than in an umbel. In composite flowers of the highest type there is something more than the mere arrangement of the florets in a compact head. The florets closely fitted together stand upon their united receptacles surrounded by one or more series of protective bracts. The outer row of flowers (rays) usually have large corollas, the petals of which, instead of forming tubes, are united to make flat signal banners more than three times as broad as a tube would be formed of the same petals. The ring of rays contrasting in color with the disk florets makes a target-like signal easily distinguished at a distance. The disk corollas are small nectar cups. The calyx tube, not being needed to protect in the usual way, becomes the chief coat of the solitary seed, while the lobes, clinging to animals or flying with the wind, in due time bear away the ripe akene. Composite flowers, then, furnishing the highest type of capitate inflorescence, stand at the head of all.

Bring for the next exercise all kinds of pods, akenes of composite flowers, and green juicy fruits.

EXERCISE 51.

Fruit and Its Use to the Plant.—Ovules fertilized by pollen become seeds, and the ovary in which they grow and ripen develops into a pod, a bur, a grain, a nut, an akene, or a juicy fruit. All these seed bearers are called fruits by botanists; but only the last—the juicy form —is known as fruit in the markets. Botanically considered a fruit is the ripened ovary, or set of ovaries, and all that directly belongs to it. Seed vessels and their attachments are fruits. Inferior ovaries become fruits in which the calyx, as in muskmelons, is a large part.

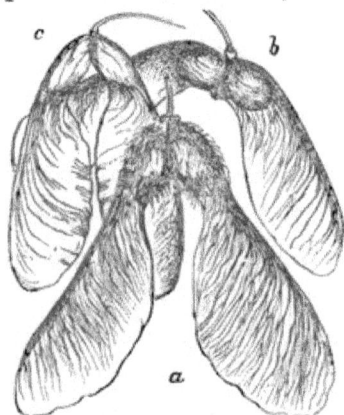

Fig. 78. Two-carpeled winged fruits (samaras). *a.* Big-leaf maple (Acer macrophyllum). *b.* Vine maple (A. circinatum). *c.* California box-elder.

Sometimes the receptacle encloses the distinct akenes of a single flower, as in roses, or the fruits of numerous flowers, as in figs. The juicy receptacle of a strawberry bears many akenes, which, like those of a fig, are commonly supposed to be seeds. Mulberries, osage

oranges, and bread fruits are developed from globular bunches of flowers, all parts of which, even the bracts, help to make up the juicy or starchy fruit. A pineapple is a similar development of a dense band of flowers, which grow below the leafy summit of the plant. Dandelion parachutes and thistle-down may be considered part of the seed-like fruits which they bear in the wind.

In studying fruits it must be constantly kept in mind that the chief use of a seed vessel is to distribute seeds, not to protect them.* The stiff, hooked hairs

Fig. 79. Akene of the coast mountain mahogany, partly enclosed by the calyx-tube.

on burclover pods, and hundreds of other fruits, enable them to get free rides to fields as yet unplanted by their kind. The styles of clematis and mountain mahogany become feathery tails; the twin carpels of maple put forth wings, and the calyx limbs of lettuce spread like silken parachutes, that the seeds may go with the wind to new homes.

Classify your fruits by separating (1) those which have no apparent means of transportation from (2) those which have. Then divide the last into (a) those which get free rides and (b) those that pay their fare.

Which of the former are carried by the wind? Which by animals? What kind of animals do they cling to? How

* A notable exception occurs in a kind of dwarf mountain pine of this coast, the cones of which usually remain closed until a forest fire leaves only the charred trunks, bearing whorls of cones nearly to the base. These, some of them more than twenty years old, soon open and reseed the ground. The fruit coats of akenes and grains are overcoats for the solitary seeds, thus doing double duty.

do they hold on ?   How do they let go?   What kind of animals carry most of the latter?   How do sheep get clover burs in their wool?   Find out by experimenting (with a feather duster for example) whether burs probably cling to birds. Does the clover-plant ever pay for the transportation of its burs?*   Did you ever involuntarily carry the fruits of wild oats, foxtail, or filaree ?   How do these fruits hold fast?   Do birds usually swallow cherry-pits ?   Would they be likely to swallow them if the pulp were much smaller?   Do you know whether coons or bears eat wild cherries ?

If there is time, draw the most interesting of your fruits.

---

* A little bird—a goldfinch, often called the California canary—is well paid for distributing the seeds of a tall and very common roadside tarweed (*Madia sativa*). The plant belongs to the sunflower family, and has yellow terminal flowers less than an inch broad, which give place to a fluted and scalloped cup a third or quarter of an inch in height and diameter. Twenty or thirty akenes, about as large as flaxseeds, but slender and nearly black, are tightly packed in this erect cup, while a dozen more, each folded in a tar-covered bract, form a loose collar around the outside at the base. The bird, in order to get the coveted nutty-flavored seeds, grasps the stem an inch or more below the cup with one foot, while the other clasps the circlet of sticky bracts. In this position he daintily picks the akenes out one at a time, cracks them, and swallows the seeds. Finishing his meal he flies away, carrying with him, sticking to his foot, the tarry bracts and the inclosed akenes, to be removed and dropped far from the mother plant. Thus, a third of the seeds are scattered at a cost of two-thirds of the crop.

Many fruits which are not transported by wind or animals have seeds which, upon the opening of the pod, are thus carried. The seeds of willows, silkweeds, epilobiums, and zauschneria have hairy tufts which cause them to travel in the wind. Pods of the castor-bean open explosively, throwing the seeds twenty or thirty feet. The seeds of some of our chaparral shrubs are thrown in a similar way. In an entirely different way the viscid seeds of pine mistletoe are shot from the contracting elastic coat of the fruit, so that they strike and stick fast to the branches of trees many feet away. The yellow oxalis fires off its smooth seeds as one may a plum-pit by squeezing it between thumb and finger. Even sweet-peas throw their seeds several feet. Indeed, there are so many devices for the distribution of seeds that students should discover some of them.

Bring for the next exercise almonds and other stone fruits (apricots, peaches, nectarines, etc.). Fruit partly grown to nearly ripe will best enable you to learn what you do not already know. Pits of the ripe fruits and husked almonds are desirable.*

EXERCISE 52.

Study of Almonds and Similar Fruits.—You could imitate an almond fruit very well by cutting a piece of thick cloth so that folded around the nut and sewed along the inturned edges it would fit it as the husk does. It is not unlike a well-filled or puffed-up turn-over pie. Indeed, it is not difficult to see evidence of the fact that almonds and all stone fruits are each made of a single leaf folded upward along the midrib with the inturned edges united so as to enclose a space which is filled by one, or rarely, two seeds. Evidently the inner surface of the shell or pit must be the upper surface of the carpellary leaf; and the skin of the fruit must be the skin of the under side of the leaf.

Note the difference between the mid-rib edge (*dorsal*) and the *ventral* edge formed by the edges of the carpellary leaf. Compare with the peach, apricot, and other fruits which you have. Which of these is most like the almond? Are the sides quite equal? Where does the

---

*If the class is working in September only peaches and possibly almonds can be easily obtained green; but the pits of other fruits are then abundant. In April, May, and June there is no want of material.

The development of these and other fruits from the condition in the flower can be profitably followed, as indicated in Exercises 21 (note) and 29.

Dried fruits soaked twenty-four hours regain their original shape.

husk of an almond open? Why should a 1-carpelled fruit open along the dorsal or ventral edges (*sutures*), or both, rather than anywhere else? Which suture do you open in shelling peas? Along which edge do the peas grow? Take a young almond and look for the stem which attaches the seed. Look for traces of the ovule which failed to grow. Why should the least number of ovules in one carpel be two? In these fruits why should only one grow? Why has one ovule become obsolete in most akenes? Compare the pits of your fruits. Draw sections similar to those in Fig. 80. In the ripe fruit you cannot make sections through the pit, but by making measurements you ought to be able to make correct sectional views. These drawings should be maps rather than pictures.

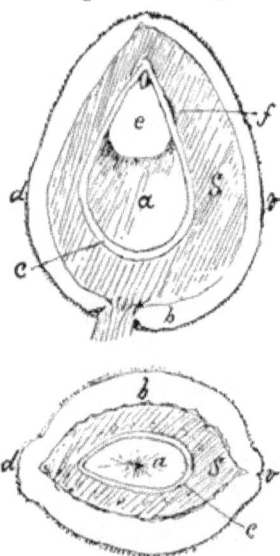

Fig. 80. Diagram of a young but full grown almond. *a.* Transparent jelly-like substance. *b.* Green husk. *c.* Skin of the seed. *d* and *v.* Dorsal and ventral sutures, or midrib and united edges of the carpellary leaf. *e.* The growing embryo. *f.* Stem of the seed. *S.* The shell of the almond.

See that they show all important facts. Get several apples, ripe or green, for the next work.

EXERCISE 53.

Study of an Apple.—Perhaps you remember that apple-blossoms have inferior ovaries. What are the leaf-like appendages at the "blossom end" of an apple? Look for stamens. Why say "blossom end " for an apple? Why not thus name the upper (*distal*) end of an almond? Cut an apple vertically so as to exactly split in halves one of the calyx-lobes; or, if there are bumps at the distal end, so as to divide one and go between two on the opposite side. Make the cut carefully down through the center, finally splitting the stem. The halves thus made should show one cell of the core—the cut passing between two on the opposite side—and a greenish line running from base to distal end. Try again if your section does not show these. Note how many seeds in the cell. Do they grow fast to the inner angle of the cell or to the outer angle? Draw a diagram showing all the facts. Put the pieces together and make a cross-section at the largest circumference. It should show a faint line outside the cells, and green dots. Draw this section, indicating the points of attachment of the seeds, the green dots and the line. Cut vertically through the dots in succession, observing carefully the sections made. What relation do you discover between dots in cross-sections and lines in vertical sections? What are these lines? Taste the pulp of a ripe apple inside the circular line shown in the cross-section. Taste the pulp outside that line. What flower leaves make the former? What do you think make the latter?*

* Occasionally an abnormal apple bears leaf blades on its surface. If there are several they are spirally arranged, as are leaves on a stem. This would seem to indicate, as is doubtless the case in figs (see Fig. 81), that the outside of the fruit is a stem or hollow receptacle; but only the blades of the leaves appear, and a thread of woody fiber, a continuation of the midrib, runs down through the pulp to the base, which favors the hypothesis that the outer and major part of an apple is made up of the petioles of the five calyx leaves.

Pupils should have perfectly formed apples to dissect. If pears are easily obtained they might be studied and compared with apples. They should be rather green. Often the fruits of roses can be easily obtained. They are interesting in this connection because they resemble apples; but a careful study shows that the apparent seeds are akenes, and that the pulp is clearly a receptacle.

Toyon berries, (very absurdly called holly), which are commonly used for holiday decorations in winter, are in condition for study from October to May.

If you have access to a fig-tree, study it this evening. Observe the size, shape, and arrangement of the leaves. Note the line running around the stem at the base of each leaf. Where do the figs grow? How many sets differing in age? Cut off a useless sprout, having ready a piece of glass or porcelain on which to catch and dry the juice.

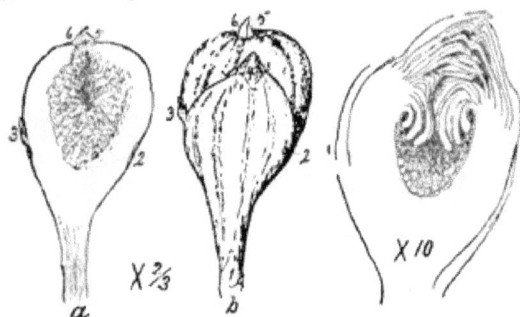

Secure two or three of each set of figs and bring them with the sprout for the next exercise.

EXERCISE 54.

Study of a Fig. —The India-rubber tree of India is a species of fig, though very unlike a fig in appearance.* Do

Fig. 81. *a* and *b*. An abnormal fig about two-thirds grown; *a* being a sectional view showing the inside filled with the real fruits, each one made up of the entire flower, peduncle, sepals and pistil gorged with sweet pulp. *b*. The outside appearance. The bracts, which are arranged spirally, are numbered from the base up. *c*. Section of a very young fig, (about one-tenth of an inch in diameter), showing the numerous bracts and the receptacle covered with blossom buds.

you think there is India-rubber in the juice of the common fig-tree?

Compare the leaves of your specimen with those of others in the class. Are the leaves of different varieties of figs alike? Look for a hole in the distal end of your fig. Note the bracts at the base. Cut a half-grown or younger fig vertically. Examine with a lens the tiny flowers which

---

* The trees of South America and Africa, which furnish most of the India-rubber of commerce, are not near relations of figs. There is India-rubber in the juice of the opposite-leaved milk weeds of this country.

Find for the next work a radish in flower and fruit. Note facts concerning size, manner of growth, size and appearance of root, condition of the soil. etc.   Bring for school study a branch with fruit and flowers.

EXERCISE 56.

Study of the Common Radish.—Wild radishes are descendants of the European species commonly cultivated in our gardens. They are all of one species, though the flowers differ in color, and the roots vary in size, color, and shape. In common with lettuce and many other plants, radishes have a great many leaves near the ground. These

Fig. 83. Study of the common radish. Made by Helen Swett, while a student at Stanford University   *a*. Base and top of the plant, one-third natural size.   *b*. Plan of the flower.   *c*. Flower seen from above.   *d*. The stamens.   *e*. The pistil.   *f*. Views of an anther enlarged.

for a time nourish the rapid growth of the tap root, which later gives up its store of food to the stem.

Examine the flowers. What kind of inflorescence? Is it bracteate. Are the sepals just alike? Are they exactly in a whorl? Remove the sepals and compare with the drawings representing them in Fig. 83. Which sepals have each a stamen opposite to them; that is, exactly between them and the pistil? Are the spaces between the claws of the petals equal? Remove the petals. Which stamens are outside or stand lower on the receptacle than the others? Are the anthers extrorse or introrse? Can you find the parts represented by two dark spots on the plan of the flower shown in the figure? What flowers have you examined which have a similar plan? Do you consider it a sign of close relationship if the flowers of two plants have the same plan? Which are of the most importance in classifying plants, the characters of the roots, or of the leaves?

Hold one of the leaves so that the light shines through it. Compare its veins with those shown in the drawing. Leaves having such veins are said to be *netted-veined*. All our native trees, except palms and yuccas, and most herbs, have such venation. When the leaves are netted-veined we expect to find the stem with pith surrounded by more or less of woody fiber; the flowers with their parts not all in threes, and seeds containing embryos with two cotyledons. These are characteristics of *exogens*.

Make a thin cross-section of the stem. Draw it. If

you have not already made drawings of the embryo do so now. Look again at the figure in Exercise 42.

Bring for the next exercise an entire plant with the flower parts in threes.

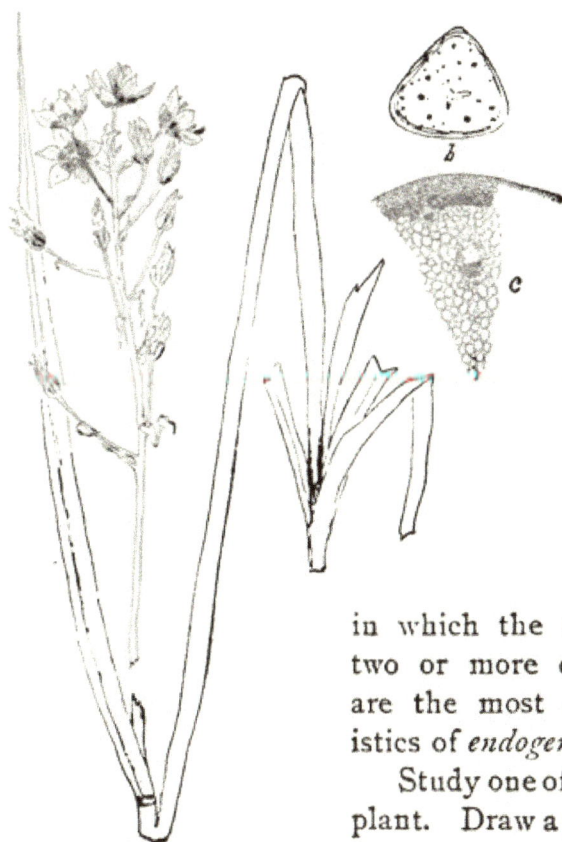

EXERCISE 57.

Study of an Endogen—Plants which have their floral organs in sets divisible by three usually have leaves with parallel veins; stems without a distinct pith; woody fibers not forming a cylinder, and seeds in which the embryos have not two or more cotyledons. These are the most obvious characteristics of *endogens*.

Study one of the flowers of your plant. Draw a plan of it and make drawings of its organs. Draw a section of the stem; a leaf or part of it; the bulb, corm, root stock, or root.

Fig. 84. Zygadenus (an endogen). Reduced from a drawing by Helen Swett. *a.* Base and top of the plant (bulb omitted), one-half natural size. *b.* Section of the stem magnified. *c.* Part of the same highly magnified.

Compare your drawings with those in Fig. 84. Note the arrangement of the woody fibers. Where are they most abundant? Which most often have long slender leaves, exogens or endogens.

All grasses, lilies, and palms are endogens. Cocoanuts, dates, pineapples, bananas, rice, corn, and all the so-called grains, except buckwheat, are fruits of endogens.

Is the calla an exogen or an endogen? If its ancestors had petals, how many were there?*

---

* Plants may be studied in this way as long as the teacher thinks best. Usually small plants like oxalis, violets, etc., should be chosen. Two or three exercises may well be devoted to one plant.

There should be review lessons based upon the note-books of the class. This important work is left in the hands of the teacher.

APPENDIX.

NOTE TO FIG. 28.

In the original pen drawing of an end view of one-quarter of a log there were faintly shown concentric arcs to represent the layers of growth. These did not "take" in the photo-engraving. With a pencil and the common attachment for drawing arcs these lines may be restored by the pupil. But it would be a good plan to have a pupil draw on the board a similar figure representing, natural size, a log six feet in diameter. The boards one inch thick should be separated by lines a quarter of an inch wide, representing the saw-kerf, or that part of the log cut into sawdust. Fine concentric lines a quarter of an inch apart would represent the growths well enough. Lines narrower than those separating the boards should be drawn at right angles to them, four or five inches apart, to mark the saw-kerfs of the saw used in "ripping" the boards into flooring. In practise, however, the log is usually not first sawn into wide boards, and these ripped; but that makes no difference with our problem, which is to understand the relation between surfaces of boards and the growth-layers. With the diagram and two pieces of real boards, representing the extremes shown at *a* and *b*, in Fig. 28, many practical questions relating to lumber can be decided, and there is no better educational work. Three or four exercises may well be devoted to it. Fine lines radiating from the center would represent the medullary rays, which are prominent in oak, causing the peculiar grain shown on the surface of boards which are "cut with the grain." The significance of the commercial term "quartered oak" is evident. The diagram shows plainly the difference between lumber from the center of a tree and that from near the surface. The following and many more questions should arise :

What part of a log will give "clear-stuff" (boards without knots)? Which furnishes the most clear-stuff, large or small logs? Explain. How many years older is one edge of your desk than the other? How do you determine? Where is the oldest wood in a tree? The youngest? Is a knot which pierces a board older or younger than the remainder of the board? Explain. What causes the wavy lines on some of the boards used in making your desk? Find the oldest wood in one of these boards. Where was the center of the tree from which it was cut? Does redwood split easier along the layers of growth or along the medullary layers? It might split more easily through the center and yet the adhesion between the seasonal growths be less than that between layers separated by medullary rays. Can you explain why?

NOTE-BOOKS AND PENCILS.

Pupils should have good pencils and note-books. Notes taken with a cheap pencil on inferior paper enclosed by paper covers are likely to be as worthless as the materials used in recording them. Good tools and good material spur the worker on to do his best. It is desirable that the leaves of a note-book be easily taken out. This will enable the teacher to take up single sheets for examination and also make it possible to rearrange them so as to bring the notes relating to one subject together. Only one side of the paper should be used. Every entry should be dated. As a rule nothing but facts discovered by the pupil should appear in the notes. The drawings should be made from the object, not copied. These and the diagrams must clearly show important facts. When it is not expedient to make the drawings natural size the fact should be indicated as shown in the drawings used in this book. For example: if the object is made twice as long and broad as natural put beside it " × 2 "; if it is only half the natural dimensions mark it " × ½." Care should be used in placing the drawings upon the page. If the working hour is short only drawings should be made in the class. These may be made with India-ink and a fine pen, or the pencil drawings, after inspection by the teacher, may be inked over. The notes may be written up at home. Only one side of the paper should be used.

The publishers furnish a good note-book. Stenographers' pencils marked M. S. are good for note-book work.

Germination Jars.

A more convenient jar than the kinds commonly used for canning fruit is furnished by the publishers of this book. It is of clearer glass and has a broader mouth. The nickel-plated cover does not fit air tight, but that is not necessary. To prevent molding the jars may be treated in this way: Fill one jar two-thirds full with water, leaving the stick in; add a teaspoonful of carbolic acid; put on the cover and shake; pour into the next jar and so on. Use but one cover in treating the jars and dry it when through. It would be well for the school to own a set of jars.

Reference Books and Books for the Teacher.

The following books should be in the school library, along with others that might well be there:

"Natural History of Plants," by Kerner and Oliver; "Flowers, Fruits and Leaves," by Lubbock; "Introduction to Botany," by Volney M. Spalding; "Elements of Botany," by J. Y. Bergen; "Structural and Systematic Botany," by D. H. Campbell; "Laboratory Practice for Beginners in Botany," by Wm. A. Setchell.

The last four are inexpensive and are worth more to teacher and pupil than all the expensive manuals and text books put together. They are so unlike that all are needed. The first book named is expensive but worth much more than its cost. It has been criticized for defense of what conservative scientists call fanciful theories or hypotheses, but it must be remembered that probably all our theories are but long-lived hypotheses. Grant White is still more "fanciful," yet worth reading by the teacher who needs inspiration.

Miss Jane Newell's Botanical Readers may profitably be used to vary the exercises.

MICROSCOPES.

Teachers should, if possible, secure for the use of their pupils the simple dissecting microscope invented by Prof. Barnes and supplied by the publishers of this work. There should be one for each pupil, but half as many are better than none. Perhaps the teacher will be obliged to get one to show the trustees and others in power before he can satisfy them of its utility. A lens of one and one-half inch focal distance is the most useful. Four or five lenses can be used together on one stand, so as to give a power of twenty diameters or more, which shows cellular structure, pollen grains, and circulation of blood in a frog's foot very well. Ten of these microscopes cost about as much as one good compound microscope and are worth considerably more than ten times as much to the pupils. Dissecting needles are indispensable, and can be made by putting handles on No. 2 sewing needles. Strong pincers will enable one to do this. One needle should be ground flat to use in cutting.

A closed cupboard is necessary for the safe storage of the microscopes, jars, etc. The shelves should be a foot wide and ten inches apart. Each set of apparatus will require about a foot shelf space. These spaces might be separated by partitions, making pigeon holes for each pupil.

DESCRIPTIVE PRICE LIST OF APPARATUS, BOOKS AND
SEEDS WHICH THE PUBLISHERS ARE PRE-
PARED TO FURNISH TO SCHOOLS.

Germination Jars with Stick, as described on a preceding page.. $0 50
Package of Seeds—the ten kinds named in the first exercise;
    enough for twenty pupils...................   50
Oil Stone...   25
Shoe Knife...   20
Liquid India Ink..................   25
Laboratory Note-book...............   25
Barnes' Microscope, as here shown, with stage plate ruled to
    decimeters and one lens of one and one-half inch focus.... 2 50
Dissecting Needles...   15
Tweezers.......................   20

INDEX.

Index.

* 9 7 8 3 3 3 7 1 0 6 1 8 8 *